STO

ACPL ITEM
DISCARDED

Y0-AAD-748

Optoelectronics

By Vaughn D. Martin

Optoelectronics

By Vaughn D. Martin

VOLUME 2
Intermediate Study
A Self-Teaching Text, Including:
- Radiometrics
- Color CRTs
- Projects

A Division of Howard W. Sams & Company
A Bell Atlantic Company
Indianapolis, IN

©1997 by Howard W. Sams & Company

PROMPT© Publications is an imprint of Howard W. Sams & Company, A Bell Atlantic Company, 2647 Waterfront Parkway, E. Dr., Indianapolis, IN 46214-2041.

All rights reserved. No part of this book shall be reproduced, stored in a retrieval system, or transmitted by any means, electronic, mechanical, photocopying, recording, or otherwise, without written permission from the publisher. No patent liability is assumed with respect to the use of the information contained herein. While every precaution has been taken in the preparation of this book, the author, the publisher or seller assumes no responsibility for errors or omissions. Neither is any liability assumed for damages resulting from the use of information contained herein.

International Standard Book Number: 0-7906-1110-4

Acquisitions Editor: Candace M. Hall
Editors: Natalie F. Harris, Loretta Leisure
Assistant Editor: Pat Brady
Typesetting: Natalie Harris
Indexing: Loretta Leisure
Cover Design: Kelli Ternet, Phil Velikan
Graphics Conversion: Debra Delk, Scott Stadler, Terry Varvel

Illustrations and Other Materials: Courtesy of 3M Corporation, Centronic, Data Display Products, Dialight, EG&G Optoelectronics, ETI Inc., Gilway Technical Lamp, Hamamatsu, Hewlett-Packard, International Light, Irvine Sensors, Keithley Instruments, Peter A. Keller, Miltec Corporation, Motorola, Raytek, SUNX, Tektronix, Texas Instruments, Visual Communications Inc., Howard W. Sams & Company, and the Author.

Trademark Acknowledgments:
All product illustrations, product names and logos are trademarks of their respective manufacturers. All terms in this book that are known or suspected to be trademarks or services have been appropriately capitalized. PROMPT© Publications, Howard W. Sams & Company, and Bell Atlantic cannot attest to the accuracy of this information. Use of an illustration, term or logo in this book should not be regarded as affecting the validity of any trademark or service mark.

PRINTED IN THE UNITED STATES OF AMERICA

9 8 7 6 5 4 3 2 1

Contents

Foreword	1
Registered Trademarks and Tradenames	4

Chapter 1
UV Theory and Applications — 7

UV's Negative Side	8
UV's Harmful Effect on Tires	8
Help Was a Hard Pill to Swallow	8
America's Own Clean Air Act	9
UV's Positive Aspects	9
But What is UV?	9
Another, Special Kind of UV	9
Keeping All the "Spheres" Straight	10
Factors Affecting the Amount of UV Reaching US	10
UV Experimental Index	10
The Dobson Unit (DU)	11
What is the Ozone Layer?	11
The Dynamic Balance	11
The Effects of Harmful Ozone Destroyers	11
How UV Affects Animals	12
Identifying Ozone Destroyers	12
Natural Ozone Destroyers	12
UV-BASED PRACTICES AND PRODUCTS	13
UV-Curable Coatings	13
Beneficial UV Curing	13
UV/EB Curing	14
Instruments Which Measure UV	14
UV LIGHT DETECTORS AND SOURCES	15
UV-Based Flame Detectors	15
Critically Examining Fire	16
UV Flame Detectors	16
UV Biological Monitoring Instruments	17
UV Curing Instruments	18
Another UV Curing Instrument	19
Theory of Operation	19
Sources of Error	19
Unit Symbols	20
A UV Light Source - The Xenon Flashlamp	21
Driving UV Flashlamps	21
Non-Semiconductor UV Sensor Vacuum Photodiodes	22

More Interesting UV-Based Consumer Products	22
More Unique UV-Based Scientific Products	23
A Fluoroptic Thermometer	25
The Flouroptic Probe	26

Chapter 1 Quiz — 27

Chapter 2
IR Theory & Application: Non-Contact Temperature Measurement — 31

A SHORT HISTORY OF THE THERMOMETER AND ITS SCALES	32
The First Thermometer	32
Improved Thermometers	32
Temperature Scales	33
IR Energy	34
Emissivity	35
Electromagnetic Radiation	36
Nongray Bodies	36
Planck's Law	37
What Happens to Emitted Energy?	39
How Does An IR Sensor Work?	40
Real World IR Pyrometers	40
Specialty Pyrometers	40
Practical Concepts in IR Pyrometer Use	41
Sensor Placement	42
Reflected Energy and Background Noise	43
Response Times	43
Sources of Error	43
Viewing Through a Transparent Window	43
Calibrating IR Pyrometers	43
Case Study #1	45
Case Study #2	47
Other Trap Problems	47
Not All Traps Are Alike	48
Diesel Pyrometer Troubleshooting Examples	49
Actual Pyrometer Instruments	50
Pyrometer Software	50
IR Sensor ICs	51

Chapter 2 Quiz — 53

Chapter 3
IrDA and Bar Code IR Applications and Wrist Instruments — 55

Optical Communications and the IrDA Standard	56
IrDA's Competition	56
Implementing the IrDA	57
IR Still Competes	58
Three Representatives IrDA ICs	59

Extending the Transmission Range	60
LED Drive Circuits	60
Making Your Own IR Filter	61
SERIAL IR HARDWARE AND SOFTWARE	62
IrDA Software	62
Development IrDA Software	63
IrDA Based Hardware	64
Troubleshooting Testers	66
Other IrDA Base Products	67
Bar Code Technology	67
A History of the Bar Code	67
An Optical Scanning Wand	68
Judging System Performance	68
Special Purpose Bar Code Characters	69
Optical Properties of the Scanned Medium	70
IR Technology at Its Best, & Its Competition	71
COMPARING THE TECHNOLOGIES	71
Non-IR Based Wrist Instruments	71
The Seiko MessageWatch	72
Message Selection	73
Limited Coverage Area	74
Reception	74
The Polar Heart Rate Monitor	74
A Stylish Beeper	74
The Breitling Emergency	75
Two Specialized Wrist Instruments	75
Optical-Based Wrist Instruments	76
The Timex DataLink	76
Sending Sample Data	77
Communication Mode	77
Setting up the Transmission at the PC	77
Sending the Data to the Watch	77
Calibration	77
DataLink Software	78
Adding Entries to the Database	78
Importing Data From Other Databases	78
The 150 Series of DataLink Watches	78
THE MODEL 150 DATALINK'S THEORY OF OPERATIONS	79
Selecting the Light Detector	79
Data Characteristics and Principle of Transmission	80
The PML	80
The TCP	80
Limitations of Data Rates	81
The Casio Infraceptor	82

The Casio Wrist Remote Controller	83
IR Based Products Aiding the Physically Challenged	83
Chapter 3 Quiz	**85**

Chapter 4
Optocouplers   87

Optocoupler History	88
An Optocoupler's Components	88
An Optocoupler's Emitter	88
An Optocoupler's Photodetector	89
An Optocoupler's Response Time	89
Using Optocouplers	89
Real World Optocouplers	90
Optocoupler Data Sheets	91
Isolation	91
Insulation	91
Response Speed	92
Propagation Delay	92
Reverse Coupling	93
CTR Trade-Offs	93
DIP Isolators	95
PC Board Layouts	96
Optocoupler Variances	96
Commercially Available Optocouplers	97
Chapter 4 Quiz	**101**

Chapter 5
Phototransistors and Optointerrupters   103

What to do With a Phototransistor's Base	104
Photodetector Similarities	104
The Photon Effect in Semiconductors	105
The P-N Junction's Photo Effect	106
Color Temperature Sensitivity	107
Phototransistor Applications	107
Optointerrupters	110
Interesting Phototransistor Based Construction Projects	112
Chapter 5 Quiz	**113**

Chapter 6
Optical Triac Drivers   115

Real World Thyristors	116
IR Optically Coupled Triac Drivers	116
Driving Actual Loads	117
Solving Inrush Current	117
The dv/dt Concept	119

Contents

The MOC3011's Input Circuit	120
Applications of the MOC3011	121
Comparing Thyristor Drive Schemes Using Optocouplers	122
Chapter 6 Quiz	**125**

Chapter 7
Photoelectric Sensing — 127

Photoelectric Sensing Modes	128
Defining Common Photoelectric Sensor Terms	129
Light Sensor Technologies	131
Modulation Types	131
Laser Photoelectric Sensors	132
Laser Sensor Precautions	136
Self Diagnostics	136
Types of Photoelectric Sensor Delays	137
Applications of Unique Photoelectric Sensors	139
Real World Photoelectric Sensors	140
Chapter 7 Quiz	**143**

Chapter 8
Experimenting With Modern Optoelectronic ICs — 145

A Photodiode's Role in Light-to-Electrical Conversion	146
The TSL250/260 Series ICs	147
The TSL214 Series IC	148
The TSL230	150
A Typical Light Modulation IC	151
The TMC3637	151
An IR-Based TMC3637 Transmitter	153
An IR-Based TMC3637 Receiver	153
IR Coupled Modulated Transmission Receiver	154
Chapter 8 Quiz	**157**

Chapter 9
Projects: Experimenting with the TSL Series of Optoelectronics IC — 159

Precautions	163
Sleep Mode	164
Mode 1 - TSL230 Evaluation	164
Hardware Design	165
A Discrete TSL230 Circuit	166
PC SOFTWARE	167
System Requirements	167
Software Photometer Operation	167
Light Meter Operation	168
Heart Rate Monitor	168

Proximity Detector Application	169
Datalogger Application	170
Chapter 9 Quiz	**171**

Chapter 10
Optoelectronic Projects — 173

THE COLOR TEMPERATURE METER	174
The Anatomy of the Human Eye	174
Color Perception Within the Eye	175
Color Perception Within the Brain	175
The Ergonomics of Color	175
Color Distribution and Saturation	176
A Short History of Color	176
Unintended Discoveries and Their Consequences	176
A Unique Use of the Camera's Direct Predecessor	177
Accident is Now the Mother of Invention	177
Types of Color Film	177
Colored Film/Light Source Incompatibility	177
A Primer on Color Filters	178
The 85 Series of Correcting Filters	178
The 80 Series of Correcting Filters	178
Light Balancing Filters	178
The Mired System	178
Using the Mired System	179
CC Filters	179
The Kelvin Temperature Scale	179
The Blackbody Concept	179
Photographic Conventions Are Hard to Shed	180
How it Works	182
The Schematic Diagram	182
The LM339 Quad Open Collector Op Amp	184
The 555 Voltage Converter	184
A Unique Counter Circuit	184
Counter Circuit Modifications	184
The Inverting Summer	185
Voltage Inversion for Op Amp Bias	185
Voltage Conversion Using a 555 Timer	185
Circuit Precautions	185
The Metering Circuit	185
Advancing the Meter	185
The Voltage Controlled PWM	186
Setting Voltage Sensitivity	186
Controlling PWM Sensitivity	186
Troubleshooting	186

Contents

Circuit Modifications	187
Insufficient Gain	187
Insufficient Negative Op Amp Bias	187
Improper Meter Deflection	187
Insufficient Dual LED Brightness	188
Incorrect Voltage Out of the U4D Voltage Regulator	188
Construction	189
Construction Procedure	189
Assembly	190
Calibration	190
THE UV RADIATION MONITOR	190
THE BURGLAR BAFFLER PROJECT	191
Solving Limited Timer Life	191
Solving Limited Bulb Life	192
Solving Predictable Light Patterns	192
Operation	193
Manual Programming	194
Program Verification	195
Block Diagram	195
Input Gating	198
AC Activation	199
Power Capacity	200
Precautions	200
Construction	200
Do It Yourself	202
Initial Check Out Procedure	202
Tips for LED Alignment	203
Non-Security Applications	203
Modifications	203
THE DTMF IR CAR ALARM	203
Technical Description of the Hand-Held Controller	204
The Alarm Module	205
The Infrared Receiver	205
The DTMF Receiver	205
The Microprocessor	205
The Signal Interface Circuits	205
The Relay Drivers	206
The Command Code Entry	206
The Construction and Operational Checkout	206
Installation and Operation	210
Battery Back-Up Option	211
DTMF Generation and Detection	211
Assembling the Transmitter's Case	213
Additional Helpful Construction Hints	213

THE BINARY-TO-OCTAL OR HEX CODE CONVERTER PROJECT 214
   The Number Base Converter's Theory of Operation 215
   PC Board Making Aids 216
Chapter 10 Quiz **221**

Appendix A
An Optoelectronics Glossary **223**

Appendix B
Sources of Supply **235**

Appendix C
Answers to All Quizzes **243**

Index **247**

*This book is for my mother
and Prof. Ron Emery*

About The Author

Vaughn D. Martin is a senior electrical engineer with the Department of the Air Force. Previously he worked at Magnavox and ITT Aerospace/Optics, where he acquired his fascination with optoelectronics. He has published numerous articles in trade, amateur radio, electronic hobbyist, troubleshooting and repair, and optoelectronics magazines. He has also written several books covering a wide range of topics in the field of electronics.

Foreword

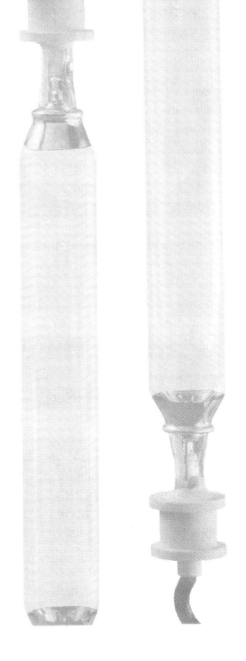

Foreword

Optoelectronics, Volume 2 continues the lessons begun in Volume 1, but also stands alone as an informative and interesting text for those individuals who are already familiar with the basics of optoelectronics, and who want to expand that knowledge. Each chapter has quizzes which allow you to verify your progress and proceed at your own desired pace. Appendix C provides the answers. This volume allows you to gain a thorough understanding of intermediate optoelectronics through "hands on" experiments and projects which have a proven track record. The experiments start from the basics with the simplest circuits possible, using even just transistors and gates. This avoids first starting with complex do-it-all ICs until you have grasped the basic concepts from these simplest possible circuits. Complex ICs, though, are still covered.

There is a kit from Texas Instruments to experiment with a sophisticated optoelectronic IC, the TSL230 light-to-frequency converter. Collectively, these tedious, time-consuming added measures make this book as easy as possible for you to affordably gain "hands on" optoelectronics experience!

Chapter 1 covers ultraviolet (UV) theory and applications, concentrating on subjects, such as dosiometers and UV-based waste water treatment. The chapter has many UV-based products, including additives to a gasoline or diesel engine and their cooling systems which brightly glow under a UV light where leaks occurs. Other practical UV subjects include flame and fire detectors, how UV affects auto tires, and fast curing UV sensitive coatings. The tutorial on ozone depletion relates to UV light by showing how bad and good ozone are molecularly identical. It's the manner in which they're generated which differentiates them.

Chapter 2 begins with a short overview of IR diodes including their structure, dark current, response speed and applications. It then investigates non-contact temperature sensing, using an IR pyrometer and explains its capabilities and also it limitations in measuring certain object's texture and color. There are several practical IR pyrometer case studies.

Chapter 3 investigates bar codes and their IR optical sensing mechanisms and various accepted bar code conventions and the 1996 IrDA standard. This standardization of transmission protocol and optical parameters allows PCs and PAs (personal assistants), such as the Apple Newton, to more interchangeably use IR based peripherals, primarily printers. This IrDA standard is now included on many of the newer HP laser printers as standard equipment.

Chapter 4 covers optocoupler applications, including the purposeful isolation of two circuits from each another, telephone circuits, signaling over distances, and techniques to enhance data rates.

Chapter 5 covers phototransistors and optointerrupters and methods to enhance their switching speeds. Applications include:

1. A light presence detector.
2. A light-driven SCR alarm.
3. An AC line voltage regulator for an incandescent bulb projector and optointerrupters.

Optointerrupters detect rates and direction of rotation by breaking a beam of light in which a circular toothed or notched wheel passes through an optical slot. These are ideal devices for detecting flywheel or rotational movements of car parts.

Foreword

Various circuits sense rotational speed and indicate when it falls outside your specified tolerances.

Chapter 6 covers the theory and applications of a specialized optoelectronics device, optical triac drivers. These isolate grounds for safety while driving 115 VAC loads. They consist of a gallium arsenide IR LED which optically excites a silicon detector IC.

Chapter 7 examines photoelectric sensing, which, for several decades simply entailed e.g., opening a grocery store's door. Now more sophisticated sensing, coupled with very affordable electronics, make possible sensing an object's size, shape, volume, distance from the source, and even its color.

Chapter 8 covers a representative sampling of modern optoelectronic ICs which detect light's presence or a specific wavelength. Their outputs typically generate a logic level after a preset light threshold activates them. There is one RF IC in this chapter solely to demonstrate you can easily convert RF into IR. The IR receiver and/or IR transmitter merely use an appropriate IR LED on their input and/or output. They supply or receive one form of energy and convert it into the other, naturally depending on which configuration you select.

Chapter 9 experiments with the Texas Instruments TSL 230 IC evaluation kit. This intelligent 8-pin mini-DIP light-to-frequency IC has a selectable 1, 10, or 100 photodiode array and a frequency output scalable —like a programmable counter IC. You'll make trade-offs in sensitivity, resolution and the number of the IC's 100 photodiodes you enable. Enabling photodiodes provides an electronic iris or the equivalent of a camera's aperture control. Enabling too many photodiodes in strong ambient light saturates the IC's light-to-frequency circuits. There is also a light intensity meter project, built with a stand alone TSL230 IC.

The TSL230 experimenter's board comes with Windows^{tm} based software, and a connector adapter for a subminiature DB-25 connector to a modular phone-style cable, allowing use with your PC's serial port. There are two pairs of matched light detectors and sources, one for visible light and the other for IR. The software creates a vivid presentation of the TSL230 connected to a counter with icon toggle switches. These provide the input voltage levels to the programming lines. On screen displays help you know which operating mode you are in and it has a real-time frequency counter display. Should you encounter a software problem, there is a quick start-up to set the baud rate, data bits, parity etc. In the photometer application you can click between autorange and manual operation and between Lux and foot candles as your light output unit symbols. There are five applications:

1. A photometer.
2. A datalogger.
3. A heart rate monitor.
4. A proximity detector.
5. An ambient light synchronization experiment.

Chapter 10 contains five optoelectronic projects. Their detailed presentation makes this longer than any other chapter. The book's more specialized chapters tend to be shorter. The first project is a color temperature meter, a handy device color photographers use to ensure a proper match of color film type, filters and ambient lighting. There are very detailed instructions for construction, test, calibration and for using the color temperature meter. There is a short history on the science of characterizing colors.

The second project is a UV radiation monitor built with just four ICs. The third project, the Burglar Baffler, is a sophisticated light activator and controller using a pseudorandom number generation scheme to activate lights between 1/8 and 7/8 into the hour(s) you've programmed to come on. It uses an optically coupled triac driver and 15 LEDs to:

1. Display a clock's face.
2. Display AM/PM indicators.
3. Display a program verification LED.

The next project is a car burglar alarm based on a wavelength matched DTMF pulse modulated IR LED sensor and detector pair. This encodes, arms and disarms the alarm. You beam the transmitter's pulses through your windshield, even highly tinted windshields! There are four detailed check out and troubleshooting charts and eight strategically placed test points to allow complete verification of proper operation and wiring.

The last project is a binary-to-octal-or-hex code converter and display, which uses just five ICs. Granted, most scientific calculators already do this. But this "hands on" effort graphically demonstrates data multiplexing, electronic selection and driving schemes of displays. You can then probe data lines to verify for yourself certain key concepts.

The three appendices provide:

1. An extensive glossary of over 175 optoelectronics terms with which you may not be totally familiar.
2. A list of optoelectronic related products and suppliers.
3. Answers to all quiz questions.

Throughout the book, commercially available products make the theory "come alive," bridging the gap between the "paper" theory of a text and more practical aspects of optoelectronics. Equations are minimally used, and only when no other means of explanation can clearly illustrate a point. Welcome to the field of Optoelectronics, an exciting technology which is useful, vitally important, rapidly emerging and constantly evolving.

Individuals from over 40 optoelectronic companies helped me in my attempt to make this book useful and practical. I'll inevitably slight someone by their omission, but thanks to the following:

Max Bernard of Hewlett Packard supplied over 50 reference publications and considerable photos and camera ready artwork.

Jack Berlien of Texas Instruments supplied a complimentary TSL230 evaluation module.

Joe Howard and Peter Keller of Tektronix.

Karen Bosco of Motorola.

Robert Angelo of International Light.

Mike Reelitz of SUNX.

John Savage of Visual Communications Co.

Gary Baker of Dialight.

Yuval Tamari of Centronic.

Joseph Blandford of Miltec Corp.

Sue Casacia of ETI Inc.

Lynn O'Mara of Irvine Sensors.

Laurie Bass of Raytek.

Registered Trademarks and Tradenames

Windows^{tm}, Windows 3.1^{tm}, Windows for Work Groups^{tm}, Windows ^{tm} and Windows 95^{tm} are registered trademarks of Microsoft.

LiteBug^{tm} is a registered trademark of International Light.

Krylon^{tm} is a registered trademark of Borden, Inc.

PAN X^{tm} is a registered trademark of Kodak.

Foreword

Cliplite™ and Cubelite™ are registered trademarks of Visual Communications Co.

Chromafilter™ is a registered trademark of Panelgraphic Corp.

Plexiglas™ is a registered trademark of Rohm and Haas.

Light Control Film™ is a registered trademark of 3M Corp.

UVtron™ is a registered trademark of Hamamatsu.

Megga-Flash™ and Lite-Pac™ are registered trademarks of EG&G Electro-Optics.

UltraViolet Sensometer™ is a registered trademark of South Seas Trading Company.

CopperTone™ is a registered trademark of Schering-Plough.

UVICURE Plus™, SpotCure™ and UVIMAP™ are registered trademarks of ETI.

SharpEye 20/20UB™ is a registered trademark of Spectrex.

Microtops™ and UV-Biometer™ are registered trademarks of Solar Light.

TouchTone™ is a registered trademark of AT&T.

Sunverter™ is a registered trademark of Abacus.

TekLumaColor™ is a registered trademark of Tektronix.

Light Control Film™ is a registered trademark of 3M.

PRISM CBI™ and Optopipe™ are registered trademarks of Dialight.

Lexan™ is a registered trademark of General Electric.

Ultralume™ is a registered trademark of Philips.

Silver Saver™ is a registered trademark of The Orchid Corp. Newton™ is a registered trademark of Apple.

Transit™ and TranXit Pro™ are registered trademarks of Puma Technologies.

LaserJet™ is a registered trademark of Hewlett Packard.

Thinkpad 755™ is a registered trademark of IBM.

THERMO-DUCER™, Thermalert™ and Raytek Field Calibration and Diagnostic Software™ are registered trademarks of Raytek Corp.

A-690 Plus™ is a registered trademark of UVP, Inc.

Chapter 1
UV Theory and Applications

Chapter 1
UV Theory and Applications

The light we use in optoelectronics resides in the IR-visible-UV electromagnetic spectrum. Ultraviolet (UV) light has the highest frequency, smallest period for one cycle, called its wavelength, and is the most "energetic" light. As an example, X-rays reside right beside UV within the electromagnetic spectrum, followed by gamma rays. Atoms or molecules absorbing light possess energy which excites their structure to "bump" up to a higher energy level. This type of excitation is wavelength dependent. Ultraviolet (UV) light is so energetic it bumps electrons into a higher orbit, but IR light just slightly vibrates atoms. This hyper-energetic UV therefore has the greatest potential for biological damage. Not surprising, UV light receives unfavorable attention for its destructive nature.

UV's Negative Side

UV causes skin aging, skin cancers, eye damage, fading in most colored man-made substances left outside and even fades car paint and degrades a car tires' side walls. These are a few examples of thousands of UV's harmful actions.

UV's Harmful Effect on Tires

Tire manufacturers blend certain chemicals, such as carbon black, into the rubber to inhibit UV's destruction. This is called a "competitive absorber" because it absorbs harmful UV, dissipating it as harmless heat, instead of combining with the tire polymer's molecules. These sacrificial molecules purposely take the UV abuse instead of the material they protect. Most tires come with a coating which naturally migrates to the surface as you drive and flex the side walls. That is why RVs and boat trailers left in the sun for extended periods without being driven or pulled dissipate their tire's carbon black, turning it gray, and "dry rot" and eventual cracking occurs.

Absorbed UV light breaks (cleaves) weak chemical bonds which cause brittleness and cracking. The bond cleavage from absorbed UV creates "radicals" which trigger a destructive chain reaction in the presence of air. This bond cleavage only requires air after it starts, even if the sun's UV light no longer exists.

Many aftermarket protective coatings are harmful to tires because they suppress this otherwise natural migration of internal waxes to the tire's surface from side wall flexing. When honoring a tire's warranty, the store often asks if you used aftermarket tire coatings. If you have, you may void many tire warranties.

Help Was a Hard Pill to Swallow

The 1987 Montreal Protocol was a promising first step toward combating ozone depletion; however, its initial rush of enthusiasm quickly surrendered to the practical constraints of compliance! It would have radically altered our lives, as we live them today. But technology is now making the war against UV and ozone depletion much easier through some novel products and procedures which collectively greatly help.

America's Own Clean Air Act

We have had national laws restricting pollution since 1955. Our present Clean Air Act of 1990 is merely an amendment of the 1963 Clean Air Act, but now taken seriously because of its vigorous enforcement. This act accomplishes six goals:

1. It establishes standards for attaining and maintaining clear air.
2. Sets emission standards.
3. Regulates hazardous air pollutants.
4. Protects stratospheric ozone and addresses acid rain.
5. Creates an air emission monitoring program.
6. Imposes strict sanctions, such as cutting off federal highway funds, against offending states.

UV's Positive Aspects

This chapter examines the causes of ozone depletion, offers alternatives to certain ozone depletion substances and practices, and profiles UV based products which contribute to UV's usefulness. UV is a germicide, disinfectant and sterilizer. Marine (salt water) aquariums support marine fish which uniquely excrete all protein liquid waste, but no solid waste, as freshwater (tropical) fish do. Surface protein skimmers help, but it's the UV which neutralizes this liquid protein waste by killing all germs. UV exposing products are now replacing traditional chemical disinfectants since UV exposure alters (severely vibrates) a cell's genetic DNA material. Bacteria, viruses, molds and algae almost instantly die.

After only 20 seconds exposure, it sterilizes the city's waste water under the UV exposing mechanism. All microorganisms die, allowing safe discharge of the treated water. This process capitalizes on UV's reaction to titanium dioxide and eliminates chlorine use, an increasingly questionable substance. UV sensor based flame detectors and UV-curable coatings on PC boards and printing inks also help reduce ozone depletion. Furniture finishing benefits from UV since fewer coatings produce the same effect, increasing production while safeguarding the ozone layer.

Other beneficial uses of UV include the new Benjamin Franklin $100 bill having color shifting dyes and security threads, visible only under UV. This makes them counterfeit-proof. Certain hobbies benefit from UV. Lapidary enthusiasts examine their minerals and rocks under UV, seeing what visible light would never reveal. UV helps stamp collectors detect repairs, errors and oddities. Model railroad signs use UV lighted fluorescent paint for a dramatic effect.

But What is UV?

UV is the sun's radiant energy and occurs in four classes or radiation bands:

1. UV-V (visible) between 395 and 445 nm.
2. UV-A between 320 and 395 nm.
3. UV-B between 280 and 320 nm.
4. UV-C between 200 and 280 nm.

Shorter UV wavelengths are more destructive (biologically damaging). UV-A is the least damaging and passes virtually unimpeded through the ozone layer. UV-B is potentially very harmful; however, the stratosphere absorbs most UV-B. UV-C is the most destructive, possessing the most energy, but the stratosphere's oxygen fortunately absorbs it.

Another, Special Kind of UV

There is another type of UV not in the UV classes previously discussed. VUV (vacuum UV) has an even shorter wavelength, higher frequency, and proportionately greater potential for biological damage. VUV has a 100 to 200 nm wavelength and is of particular interest to scientists. You will never encounter this type UV in nature because it is man-made and so readily absorbed by gases that it has to be created in a vacuum. Most small molecules have their first electronic absorption transitions in this region, thus most gases are invisible.

Special lamps produce VUV by a process called synchrotrons or by four-wave mixing of lasers in gases. As an example, you could add three low frequency photons together to produce a single high frequency photon. This is called sum frequency mixing. You can also produce a photon whose frequency the sum of two photons minus the frequency of a third. This is called difference frequency mixing. Due to a quirk in physics, the sum frequency is disallowed at wavelengths where phase matching is not met. Therefore, it tends to be difficult to make a broadly tunable light without special precautions. Difference frequency mixing fortunately has no disallowed wavelength regions and is understandably easier to broadly tune.

An ArF laser is resonant with this light transition so it does not require any tuning. The gas with which this occurs most easily is hydrogen and it does so at a few hundred Torrs pressure. Krypton is a gas which works not quite as well. VUV has experimental uses as a semiconductor substrate cleaner and etcher.

Keeping All the "Spheres" Straight

The stratosphere is from 15 to 50 kilometers above the earth. This area experiences an increase in temperature with a corresponding increase in altitude due to absorption of UV light by oxygen and ozone. This creates a temperature inversion, impeding vertical motion into and within the stratosphere. Stratosphere is a word related to stratified or "layered." The troposphere is from 15 kilometers to the earth's surface. Above the stratosphere is the mesosphere which ranges from 50 to 100 kilometers from the earth's surface. Above this, from 100 to 400 kilometers, is the exosphere which starts to reach interplanetary space.

The air immediately surrounding the earth's surface is actually the troposphere, but is nebulously also called the atmosphere—in contrast to the stratosphere. The "ozone layer" resides within the stratosphere and is concentrated at about a 25 mile or 40 kilometer altitude.

Factors Affecting the Amount of UV Reaching Us

Since the sun reaches the Earth's equator at the most direct angle, it receives the most UV. During winter the least UV reaches us and the most occurs during summer, conforming to a Bell shaped intensity curve with the Northern hemisphere experiencing its peak around the 185th Julian day of the year (midsummer). Thinner air, characteristic of higher altitudes, allows more UV rays to pass; however, added cloud cover, smog, and rain reduce UV. Land cover reflects about 85% of UV, with sand and snow reflecting 12% and water reflecting just 5% of the UV reaching it. Reflected UV damages people, plants and animals, just as direct UV does.

UV Experimental Index

In the summer of 1994, the National Weather Service began offering the UV Experimental Index as part of their weather report. This forecasts the UV reaching the earth's surface at its peak hour, noon. It also includes the dampening effect of cloud cover. In summer, the index ranges from 0 to 15. Schering-Plough, the makers of Coppertone^{tm}, are strong advocates of this index since skin cancers are growing at a rate 4.5% per year in America, but Australia's sun worshipers even top this. The EPA offers a free publication, Bulletin of Sample Public Safety Messages to Accompany the Experimental UV Index. Appendix B lists the EPA and National Weather Service's UV index programs and phone numbers.

Theoretically, the ozone layer should be the thinnest at the equator and the thickest at the Earth's poles. However, the Antarctic ozone hole was a clarion call for action. By the 1980s, scientists observed as much as a 60% ozone reduction. In October, 1992, there was an entire column over the South Pole with ozone levels lower than at any time in this area's 35 years of record keeping. This interestingly coincided with Mount Pinatubo's volcanic eruption. The surface readings were less than 100 Dobson Units (DU). This was down considerably from 321 DU in 1956. Between 13 and 18 kilometers in altitude there was a complete ozone void or absence of this UV shielding layer.

The Dobson Unit (DU)

This is a measure of the ozone layer occurring directly above the earth, in a column. It is an ozone thickness of 0.01 mm at 0° C and 1 atmosphere pressure. The average thickness of the ozone layer in America is 300 DU. Comparing industrialized versus developing areas, Huancayo, Peru averaged 255 DU in April and St. Petersburg, Russia had 425 DU.

A DU is approximately 2.7×10^{16} molecules/cm². This unit was named after G.M.B. Dobson, the pioneer in atmospheric ozone studies from 1920 to 1960. Dobson also designed the standard instrument with which you measure ozone directly above the earth. It used a spectrophotometer to measure UV radiation at four points - two of which the atmosphere absorbs and two which it doesn't absorb. But today, the more sophisticated Brewer spectrophotometer is gradually replacing it.

What is the Ozone Layer?

UV coalesces with and causes ozone depletion. The oxygen we breathe is diatomic (two atom) oxygen (O_2) and ozone is (three atom) triatomic (O_3) oxygen. Ozone is colorless but has a harsh acrid smell and 90% of it occurs in the "ozone layer" 15 miles above the earth's surface, which serves as a beneficial UV sunscreen. The other 10% of ozone is produced at ground level from a reaction between sunlight, volatile organic compounds (VOCs) and nitrogen oxides (NO_x) and is merely smog. Despite both types of ozone being molecularly identical (O_3), their occurrences at different altitudes prompt vastly different consequences. Ground level ozone absorbs a small amount of UV but is mainly smog. Smog becomes very serious though in temperature inversions or when warmer polluted air is trapped on the earth's surface by heavier colder air above. Another adverse effect of sulfur and nitrogen oxides is acid rain which is related to sulfuric and nitric acid and originates from burning fossil fuels. Acid rain occurs when dry air carrying these pollutants collides with wet air and falls to earth as rain, smog, mist or fog.

The Dynamic Balance

In an ideal unpolluted world, there is a constant quantity of ozone. Ozone's perpetual creation and destruction are equal. The highly energetic sun's rays in the stratosphere strike molecules of oxygen (O_2) and split the diatomic oxygen into atoms of single oxygen. These two resulting single oxygen atoms combine with a diatomic oxygen molecule to form two triatomic oxygen molecule (O_3), or ozone. Ozone is naturally broken down in the stratosphere in very small amounts by various compounds containing nitrogen, hydrogen and chlorine. Ozone depletion is more prevalent on our surface, with the release of chlorine and bromine from synthetic compounds (halo carbons), the main culprits.

The Effects of Harmful Ozone Destroyers

Halo carbons are effective ozone destroyers because:

1. They do not readily combine with other chemicals in reactions, rendering them harmless.
2. They also promote natural ozone-destroying reactions.

Rain does not wash halo carbons back into the atmosphere. They remain in existence from 20 to 120 years, depending upon the compound. This is ample time to drift up into the stratosphere or where they help eradicate the ozone layer.

The stratosphere breaks these highly energy charged UV-C molecules into chlorine form (CFCs, methyl chloroform, carbon tetrachloride) or bromine form (halons and methyl bromide). Both chlorine and bromine activate and hasten ozone destruction. A single chlorine atom can destroy up to 100,000 ozone molecules before it finally forms a chemically stable compound and diffuses out of the stratosphere. Dr. F. Sherwood Rowland and Dr. Mario Molina first hypothesized this chlorine and bromine ozone destruction theory and were initially met with abject skepticism. However, after the British Antarctic Survey Team measured severely declining ozone layers over Antarctica's Halley Bay in the mid 1980s, criticism turned into acceptance.

How UV Affects Animals

Another factor buttressing their theory was certain marine animals, such as small finfish, shrimp and crab larvae experienced marked changes in their developmental cycles. Sheep in the Falkland Islands off the southern East coast of South America were blinded from a greatly rarefied stratosphere's excessive UV. The ozone layer was almost completely eradicated.

Rodents have an acutely UV sensitive retinal mechanism. Birds detect UV and select a mate on the basis of color, as UV reflects off a potential mate. Human hunters are at a disadvantage who wear clothes washed in laundry detergent. They also possess a strongly UV reflective quality. UV destroys amphibian eggs — thus their dwindling numbers. Lizards reflect UV and iguanas use UV to detect edible plants in the desert and see territories staked out by other iguanas.

Identifying Ozone Destroyers

Approximately 80% of all stratospheric ozone destruction occurs from emissions of CFCs (chlorofluorocarbons). These are primarily the coolants in refrigeration, air conditioners, solvents for paint and de-greasers and blowing agents in the production of foam. Cars made before 1990 used an environmentally unfriendly air conditioner coolant, but the EPA has made manufacturers cease production since late 1993. Then, a 30 pound canister cost about $28. Today it is in excess of $700 if you are lucky enough to just find one. The alternate environmentally friendly coolant requires retrofitting a car's air conditioner, an expensive task. The following are the five most pernicious and pervasive ozone destroyers:

1. HCFCs (hydrochlorofluorcarbons) contain chlorine; however, unlike CFCs, they also contain H (hydrogen) which promotes their breakdown in the atmosphere. Collectively, they are called transitional elements since they are an interim step between strong ozone destroyers and replacement chemicals that are entirely ozone friendly.
2. Carbon tetrachloride is a substance used as fumigants by farmers and in petrochemical refineries. It accounts for less than 8% of total ozone destruction.
3. Methyl chloroform is used to clean electronic parts and metals and is a strong ozone destroyer with growing widespread use.
4. Halons are fire suppressants but due to their long lives they are growing at a rate of 10% per annum.
5. Methyl bromide is a pesticide which contributes 5 to 10% of the annual total ozone destroyed.

Natural Ozone Destroyers

Volcanoes are a dramatic and effective source of ozone destruction. As volcanoes erupt and spew both dirt and hot gases into the stratosphere the empirical observation of a plummeting concentration in the ozone layer is undeniable. El Chichon in Mexico in 1982 caused global reduction of ozone by 2%. A more powerful volcano, Mount Pinatubo in the Philippines, erupted in June, 1991, reducing local ozone levels by 20% even six months after the eruption.

UV-BASED PRACTICES AND PRODUCTS

UV-Curable Coatings

Printing and coating industries extensively use UV to dry and cure UV-curable inks and coatings. Many commercial products coated with a high quality paint or clear sealer use UV based curing. Clear hardeners on billiard balls, skis, aircraft surface paint, and furniture are a few examples. UV coatings contain three primary ingredients: polymer-based resins, monomers or cross-linked oligomers and photo initiators. UV light exposure prompts the photo initiator to start a chemical chain reaction which cross-links and polymerizes the UV sensitive coating. The alternative to UV curing is solvents which evaporate and release damaging VOCs into the atmosphere.

Beneficial UV Curing

These case studies demonstrate constructive uses of UV. A printer uses UV-curable inks too but added water-soluble acrylics for colors, discovering IR light dries them. The IR and UV light combination produced an effective process. Another printing firm uses UV-curable inks and mixes them

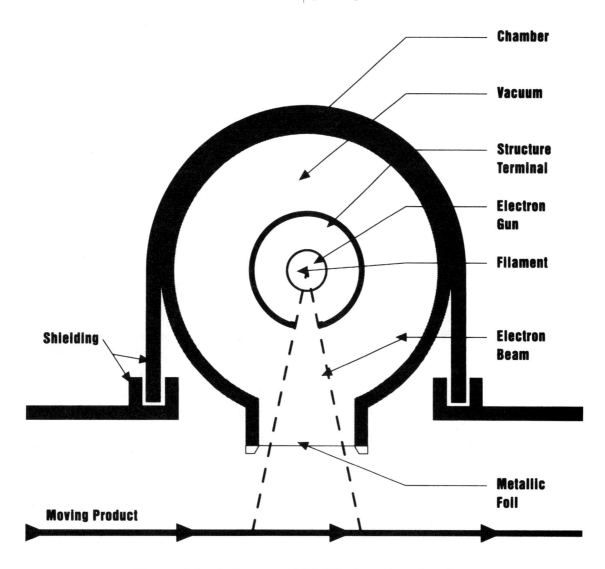

Figure 1-1. A diagram of EB (Electron Beam) curing.

with solvent-based inks, which an approved hauler disposes of legally. Subsequent jobs use any excess UV-curable inks for color mixing.

A wallet insert manufacturer formerly hand stitched products but vinyl proved useful in other areas, such as portfolios, pocket secretaries and promotional products with their obligatory company logos. But the inks in the silk screen dried slowly and consequently frequently smudged. Slowly drying and highly flammable inks with a strong outgassing smell harm the environment. UV light exposure of UV-curable inks dries them in less than a second — with no smudging!

A silk screen is typically a wooden frame holding a piece of tightly stretched silk, usually approximating a square. The treated silk does not pass inks in certain areas but easily passes inks in other areas, forming patterns, letters etc. A squeegee type window cleaning device evenly spreads the ink. Using screens with different colors of inks produce multiple colors. Many commemorative tee shirts use this method.

UV/EB Curing

This stands for UV and Electron Beam curing. The sources of UV are typically one of the following:

1. Medium pressure mercury lamps.
2. Pulsed xenon flashlamps.
3. Lasers.

You are probably familiar with all these technologies, except a pulsed or flash xenon flashlamp which we'll cover shortly.

An electron beam accelerator (See *Figure 1-1*) though cures thicker layers. *Figure 1-2* shows UV curing in which the surface tends to absorb UV light photons and not penetrate as deeply as electron beam curing.

Instruments Which Measure UV

There are many instruments which specifically measure specialized aspects of UV such as:

1. UV intensity.
2. UV power.
3. UV's biological effects on the ozone.
4. UV dosage energy which is the multiplied product of UV intensity and time.
5. How well UV cures a product.
6. UV sensor-based flame and fire detectors.

Let's start with UV-based flame detectors (sensors) and UV light sources.

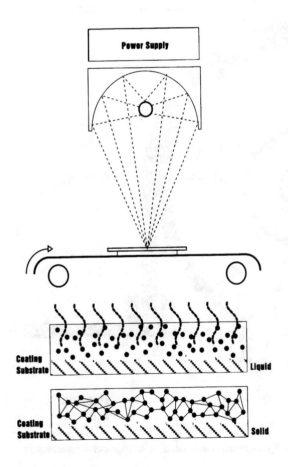

Figure 1-2. A diagram of UV curing.

UV LIGHT DETECTORS AND SOURCES

UV-Based Flame Detectors

These optical narrow band sensor-based instruments use either:

1. UV sensors.
2. IR sensors.
3. IR and UV sensors.
4. Triple IR sensors (See *Table 1-1*) to act as a flame/fire alarm.

You use UV flame detectors mainly indoors; therefore, they sometimes encounter radiation sources which induce false alarms. These include arc welding, lightning and excessive amounts of sunshine.

UV sensor-based instruments analyze fire by:

1. The fuel.
2. Fuel consumption.
3. Oxygen/air consumption.
4. Evolving heat.
5. Chemical reactions in the vaporized fuel.

Fire's radiated energy is a major data detection source. Between 30% to 40% dissipates at various wavelengths from IR to UV.

Flame detectors use optical sensors with several specific narrow spectral ranges which record the incoming data at the selected wavelengths. These

COMPARISON BETWEEN VARIOUS TYPES OF FLAME DETECTORS

TECHNOLOGY	ADVANTAGES	DISADVANTAGES	APPLICATIONS
OPTICAL DETECTORS			
INFRARED (IR)	High speed Moderate Sensitivity Manual self test Low cost	Affected by temperature Subject to false alarms IR sources	Indoors
ULTRAVIOLET (UV)	Highest speed High Sensitivity Automatic self test Low cost	Affected by UV sources Subject to false alarms Blinded by thick smoke & vapors	Indoors
DUAL DETECTOR (UV/IR)	High speed High sensitivity Low false alarm rate Automatic self test	Affected by specific UV/IR Ratio created by false stimuli Affected by thick smoke Moderate cost	Outdoors/Indoors
DUAL DETECTOR (IR/IR)	Moderate speed Moderate sensitivity Low false alarm rate	Limited operation by temperature range Affected by IR sources Moderate cost	Outdoors/Indoors
TRIPLE IR	High speed Highest sensitivity Lowest false alarm rate Automatic self test	Moderate cost	Outdoors/Indoors

Table 1-1. The proper optical technology with which to detect various types of fires. (Courtesy of Spectrex Inc.: "How to Choose the Right Detector for your Application")

instruments analyze data by one or more of the following methods:

1. Flicker frequency analysis.
2. Threshold energy signal comparison.
3. Math correlation between several signals.
4. Boolean comparison techniques of AND, OR or ratios.
5. Correlation to a ROM lookup table composed of specific spectral analyses.

The surrounding atmosphere of air, dust, smoke and gases readily absorbs the short wavelength UV. UV radiation in the range of less than 300 nm is absorbed and will not create any false alarms. UV detectors can detect a flame in less than 4 milliseconds, but use longer sampling intervals to avoid false alarms.

Figure 1-4. *Spectrex's SharpEye^{tm} 20/20UB self-testing UV flame detector. (Courtesy of Spectrex)*

Critically Examining Fire

A fire starts from its combustibles and produces by-products. A hydrocarbon based fire's spectrum peaks at a 4.3 μm wavelength, making this an obvious candidate for an IR-based fire detector. Only a small peak occurs within the UV spectrum. Examples of typical industrial applications absolutely critical for fire detectors include an aircraft hangar which typically uses four fire detectors, each lobbing approximately 25 feet to completely cover most medium sized planes. Another less obvious application is an oil storage tank with a roof which floats on the oil. This uses IR beam detectors with beams crisscrossing the roof since they are also mounted on this oil storage tank's floating roof.

UV Flame Detectors

Figure 1-3 is the model C7050 UV flame detector from Det-Tronics. Note the lens pointing outward and the circular turreting swivel flange mounting bracket. This works with the controller-based system with relays and 4 to 20 mA current loop out-

Figure 1-3. *Det-tronics' model C7050 flame detector. (Courtesy of Det-tronics)*

puts. The controller's digital display pinpoints any detector indicating a fire. This UV detector responds in 10 milliseconds; however, you typically use a 3 to 5 second integration sampling period delay to guard against false alarm triggering fast light transients, such as arc welding.

Figure 1-4 is the Spectrex SharpEye^{tm} 20/20UB, a self contained UV flame detector. This field programmable instrument allows user defined delays and multiple detection levels for pre-alarm, alarm and saturated signal responses. It also has built-in test features capable of testing lens cleanliness and proper operation of the sensors. Its military specification components yield a very reliable 100,000 hour (over 11 years) mean time between failures.

UV Biological Monitoring Instruments

Solar Light specializes in this market niche and two representative examples are their Microtops^{tm} II ozone monitor and their UV-Biometer^{tm}. You just aim the Microtops^{tm} II total ozone, water vapor and sun photometer (See *Figure 1-5*) at the sun through the sight's cross-hairs and push a button. This instrument's low noise electronics and 20 bit A/D converter produce:

1. High linearity.
2. High resolution.
3. A wide dynamic range.

This five channel sun photometer's center wavelengths are 300, 305.5, 312.5, 940 and 1.020 nm. It automatically calculates the ozone and water vapor column based on:

1. Measurements at three UV wavelengths.
2. The site's latitude and longitude.
3. The universal time.
4. Atmospheric pressure through its built in pressure transducer.

Figure 1-5. *Solar Light's Microtops^{tm} II total ozone, water vapor and sun photometer. (Courtesy of Solar Light)*

Figure 1-6. *Solar Light's model 501 UV-Biometer^{tm}. (Courtesy of Solar Light)*

Figure 1-7. *ETI's SpotCure™ UV hand-held intensity meter. (Courtesy of ETI)*

You test this instrument's performance against a Dobson spectrophotometer.

The UV-Biometer™, model 501 (See *Figure 1-6*) measures the biological effect of UV light by paralleling the steep Erythema Action curve which is common to many biological action spectra. This instrument measures 280 nm UV radiation with a combination detector of absorption filters, phosphor and a GaAsP LED. This instrument is an upgrade of the pioneering Robertson-Berger meter first used at Temple University's UV global network in 1973.

UV Curing Instruments

The ETI UVICURE™ plus is quite similar to the Light Bug™. (See *Figure 1-10*) It is also a half inch thin peak UV intensity and total exposure energy meter made to slip into an oven for UV curing applications. This instrument allows you to measure UV lamp performance and compare the efficiencies of curing systems. UV curing systems channel a high intensity UV light source through a liquid light guide which controls high intensity UV and focuses it onto the small spot to be cured. UV curing systems predominantly consist of a lamp, reflector and liquid light guide, all of which deteriorate from the severity of UV over time.

These light guides are often bent or twisted and do not precisely direct the UV at the desired spot. The ETI SpotCure™ Intensity Meter (See *Figure 1-7*) resembles a small flashlight in size and shape and has multiple waveguide adapters to allow you to use different sizes of light guides. You can use it to measure the UV output and then empirically position the light guide for optimum UV exposure on the intended target. This UV meter displays UV intensity results in Watts/cm^2, allowing you to determine:

1. Spot curing system performance.
2. UV light degradation.
3. Optimum positioning of light guide cables.
4. The efficiency of UV spot curing systems.

Figure 1-8 is ETI's UVIMAP system. The UVIMAP is the middle instrument which measures both temperature and UV radiation levels as a function of time as it rolls along a production conveyer belt. Its printer is on the left with cables, manuals and IBM PC compatible software. The instrument is in a Fiberglas case painted with Sperex, a silicone-based paint used on tail pipes of dragsters which is unaffected by UV radiation and heat. Its matte surface deflects and diffuses the incoming UV. The instrument's back has screws for adjusting scan rate and intensity range.

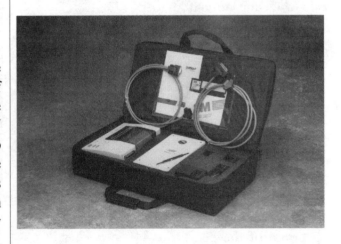

Figure 1-8. *ETI's UVIMAP™ UV detection system. (Courtesy of ETI)*

Another UV Curing Instrument

International Light's IL390B, called the "Light Bug™" (See *Figure 1-9*), measures UV doses and has a low profile (just 1/2 inch high) permitting its insertion into UV curing ovens, printed circuit (photoresist) and printing plate exposure systems. Its purposely highly polished aluminum case reflects the high IR and UV irradiance of ovens and UV curing environments. Note the small light aperture (detailed by a "hand") and the 280 nm lower limit of UV-B response curve.

You position this instrument in the same place as the object receiving UV, ensuring the same dose of UV as the treated product. A typical operating scenario is first adjust the conveyer belt speed to optimize the UV dose. Next measure UV exposure by pressing the only button on this instrument. Finally, at the start of each day send the product through the oven again, adjusting the conveyer belt speed to duplicate the UV dose at which best UV curing occurred.

Theory of Operation

The Light Bug™ allows light to enter a small aperture where it travels through a narrow slit, absorption filter and finally falls upon a vacuum photodiode. This converts light to current and is amplified again and converted to optical units the LCD displays. The instrument has a wide spectral response. This is a measure of the signal from the detector (vacuum photodiode) in response to light of constant irradiance at each measured wavelength (UV in our case) changes. This instrument has a broad UV range from 250 to 400 nm. This accommodates the spectral sensitivities of a wide variety of UV curable materials, including UV-V. These photocurable monomers each have their own unique photochemical spectral sensitivities. But they must most be concerned with the UV curing spectrum of the photoinitiator (absorbing dye) which absorbs the UV to produce cross linking and polymerization.

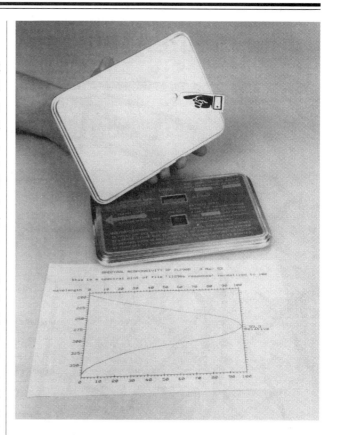

Figure 1-9. *International Light's Light Bug™ UV dosiometer showing both sides, its sensitivity curve and aperture — see the small pointing hand. (Courtesy of International Light)*

Sources of Error

Ideally, the spatial response corresponds to a Lambertian cosine receiver. That is, the relative spatial response of a flat surface equals the cosine of the off-axis angle of the light source. Therefore, if the light source is at 45° to a flat 1 cm^2 surface, it receives only 0.707 cm^2 of light. *Figure 1-10* shows how closely the Light Bug™ conforms to the ideal Lambertian cosine receiver. The narrower the photodiode detector's spatial response is, the more erroneous the measurement. This instrument uses diffusing material to ensure accuracy without degrading sensitivity.

The vacuum photodiode experiences a change in sensitivity of 0.2% per °C. There is a miniature integrating sphere within the Light Bug™ which

RELATIVE SPATIAL SENSITIVITY

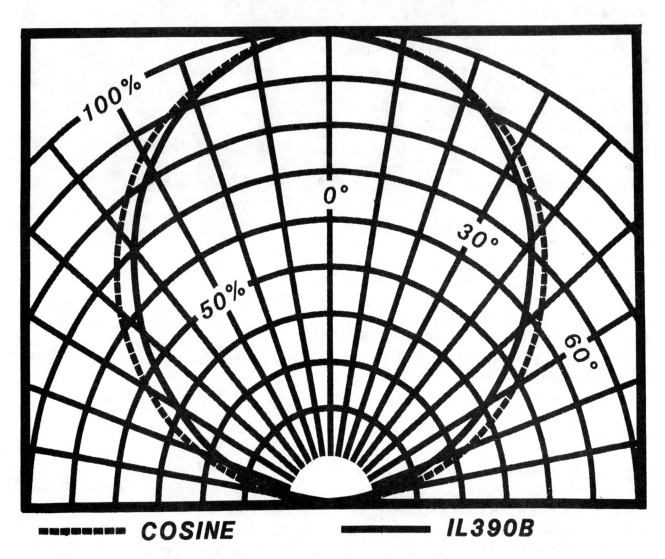

Figure 1-10. *The Light Bug™'s performance against an ideal Lambertian cosine receiver. (Courtesy of International Light)*

prevents reflective degradation due to prolonged exposure to high temperatures. The sphere is a pressed barium sulfate cube. This provides high UV reflectance just as barium sulfate paint does; however, it does not break down as microscopic impurities from paint volatiles do.

Unit Symbols

The Light Bug™ measures the UV exposure in energy per unit area (joules/cm²). If you require power per unit area (Watts/cm²) you can use the following equation to convert to these units.

$$1 \text{ joule/cm}^2 = 1 \text{ Watt/c}\bullet^2 \text{ x exposure time}$$

Energy is measured in joules and is what does work. The rate of work is energy per unit time. A work rate of one joule per second is one Watt. We do not use joules and Watts alone when accessing long UV curing exposure times since the cured product acts as an extended receiver. The Light Bug™ collects all light incident upon it and divides

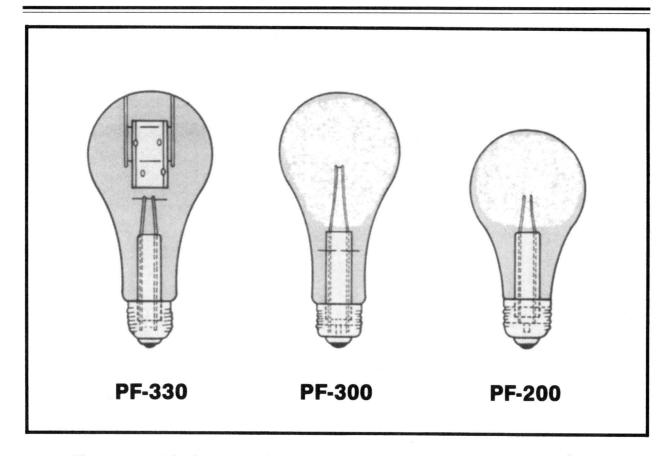

Figure 1-11. EG&G's Electro Optics' Megga-Flash™ UV series of xenon flash lamps. (Courtesy of EG&G)

it by the detector's area in cm^2. It integrates or averages this over the exposure time in seconds which equals energy per unit area (joules/cm^2).

This is typically 0.5 to 5 joules/cm^2. The Light Bug™ measures directly in millijoules/cm^2 to allow for finer resolution on the display. To convert the displayed reading to milliwatts/cm^2, divide the reading by the exposure time. Since integration times are typically 30 seconds to 3 minutes, the detector's response time is a negligible source of error.

A UV Light Source - The Xenon Flashlamp

The Megga-Flash™ is a representative UV flashlamp light source from EG&G Electro-Optics. (See *Figure 1-11*) These bulbs range in height from 4.75 to almost 6 inches; *Table 1-2* lists their salient characteristics. High speed motion picture photography, large format photography, destructive testing, and curing applications use them. They produce literally millions of Lumens of light in milliseconds. (See *Figure 1-12*) They do have a strong spectral peaking in the UV-C region, obviously that's why destructive testing uses them. *Table 1-3* lists this flashlamp's spectral content by percentages.

Driving UV Flashlamps

Figure 1-13 shows a driver for smaller one inch tubes, EG&G's Lite-Pac™. Note how high the main discharge voltage is, up to 1500 VDC! Since this pulsed device releases great amounts of quickly fleeting energy, it uses SCR triggering. The illustration's equations define the operating parameters but predict peak currents which ignore the in-line circuit resistance. Also, you only need the CR diode when you are not using a Lite-Pac since it already has an internal diode.

	EG&G Part Number		
	PF-330	PF-300	PF-200
Class	Flood Flash	Slow Peak	Medium Peak
Duration	1.75 seconds	– –	– –
Lag Time	50 msec	– –	– –
Peaking Time	– –	30 msec	20 msec
Rated Lumen Seconds	140,000	110,000	70,000
Approx Color Temp	3800°K	3800°K	3800°K
Min BC Power	10 MWS	10 MWS	10 MWS
Bulb Type	A-23	A-23	A-19
Base Type	Med screw #102	Med screw #102	Med screw #102
Voltage	4.5-45 volts	3-125 volts	3-125 volts
Amp	3	3	3
Replaces Sylvania Bulb	Type FF 33	Type 3	Type 2

Table 1-2. EG&G's UV Megga-Flash^{tm}'s salient characteristics. (Courtesy of EG&G Optoelectronics)

Non-Semiconductor UV Sensor Vacuum Photodiodes

These flame/fire detectors use various types of UV sensors. The Hamamatsu UVtron^{tm} UV flame and fire detecting non-semiconductor vacuum photodiode is also well suited for detecting discharge phenomena such as the invisible corona discharge of high voltage transmission lines. It can detect a cigarette lighter in a room at greater than 15 feet by using its metal's photoelectric effect and the gas multiplication effect. This produces a narrow sensitivity band of 185 to 260 nm. This is the most destructive type of UV, UV-C light. This UV photodiode is totally insensitive to visible light and therefore requires no optical filters. Since UVtron^{tm} photodiodes also emit UV, exercise care not to place them too close together or they will optically interference with each other.

More Interesting UV-Based Consumer Products

There is a product for just $5 which is the size and shape of a credit card. It is the UltraViolet Sensometer^{tm} from South Seas Trading Company in Hawaii. It measures UV strength in about 10 seconds to determine the ultimate sunscreen tan-

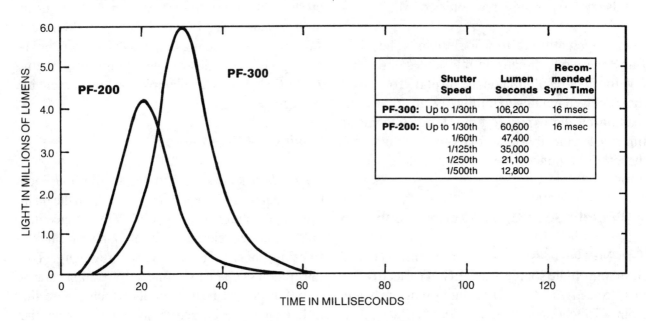

Figure 1-12. The Megga-Flash^{tm}'s Lumens output versus time. (Courtesy of EG&G)

Chapter 1

Light Output Distribution

Wavelength Intervals (nanometers)	Percentage Standard Bulb	UV Bulb
200– 300	—	18
300– 400	11	14
400– 500	28	22
500– 600	15	12
600– 700	12	7
700– 800	9	6
800– 900	11	8
900–1000	8	7
1000–1100	6	6

Standard Bulb (Standard Glass)

Table 1-3. EG&G's Megga-Flash™'s percentages of spectral output. (Courtesy of EG&G Optoelectronics)

ning lotion to use. The UV sensor strip at the card's bottom turns color and each color uniquely identifies the level of UV intensity. If you place the lotion on the card itself and expose it for 10 seconds you will be able to tell when it wears off, indicating you need to reapply more protective lotion to your body. The card checks if your PC or TV is emitting excessive UV and the UV screening effectiveness of your sunglasses. Hold them over a portion of the card's sensor strip in the sun. If your sunglasses have true UV protection, the sensor strip remains white under the sunglasses and turns blue in unprotected regions.

UPV produces a leak detection system for auto, boat and heavy industrial diesel engines. These fluorescent fluids allow you to spot leaks as a bright fluorescent glow under UV. You can also add UVP's A-690 Plus to your gasoline engine, then run it to circulate this UV sensitive additive.

More Unique UV-Based Scientific Products

The PixelVision UV-sensitive digital CCD technology produces very UV-sensitive cameras. These 16-bit CCD technology cameras offer excellent sensitivity and greater than a 45% quantum efficiency. Unlike most visible CCDs, which lack sensitivity in the blue region, PixelVision's UV and anti-reflection coated CCDs are available in both line scan and array formats.

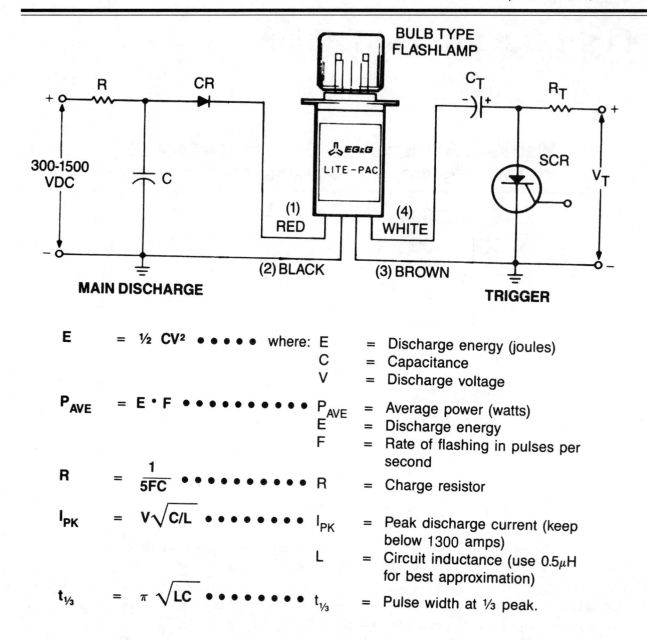

Figure 1-13. EG&G's Lite-Pac^{tm} for driving smaller xenon UV bulbs. (Courtesy of EG&G)

Interestingly, PixelVision is best known for a product which was a big flop. They made the PXL-2000 video camera for Fisher-Price toys in the late 1980s. They quickly took this off the market; however, a few years later it began showing up at film festivals. Today, clubs with cult like followings trade tapes, spare parts, and generally covet this camera. Its distinctive feature was that it recorded both sound and video on ordinary tape, but due to its limited bandwidth, produced an image resembling black and white if it were left in the middle of a busy street and then run through a projector. Its grainy image, considerable dropout and weird slow motion appearance produce a surrealistic effect. These normally objectionable characteristics strangely contribute to its appeal. There is a three minute bar scene in the 1992 film Slacker by Richard Linklater shot by this camera. In the 1996 film about a New York City vampire, Naja, the camera shows what the world looks like through the eyes of Dracula's daughter.

Chapter 1

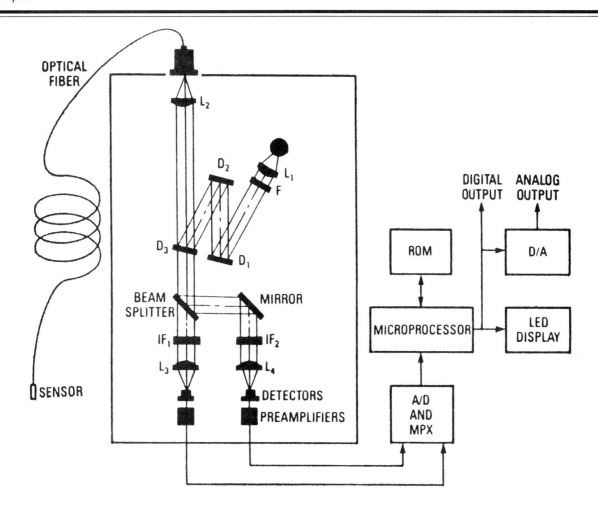

Figure 1-14. A block diagram of a fluoroptic UV-based thermometer.

Sensor Physics recently released the first version of its Windows™ 3.1 or 95 based software for UV beam analysis. In addition to beam parameters, the software includes a user definable lookup table for absolute dosimetry. Sensor Physics' UV Sensor Cards allow instant, high resolution measurement of both CW and pulsed UV lasers from 500 to 8 Ångstroms.

Oriel Instruments has solved a vexing problem with UV light sources. This is maintaining a constant intensity of light output regardless of aging, ambient temperature, or line voltage variations. These UV intensity controllers have two primary components, a light sensing head and a controller unit. The light sensing head, which houses a UV enhanced temperature stabilized silicon photodiode, monitors part of the light output. The controller constantly compares the recorded signal with the set level, and if a variations occurs, it adjusts the power supply to the desired UV intensity. Incidentally, standard glass absorbs wavelengths less than 300 nm. For UV detection, you will use a fused silica or a special UV transmitting glass.

A Fluoroptic Thermometer

This non-contact temperature measuring IR pyrometer requires a direct line-of-sight path for temperature measurement. This renders IR sensor-based technology useless for hard to access places. The Luxtron company invented a fluoroptic thermometer. (See *Figure 1-14*) This device senses temperature by using slender optical fibers, per-

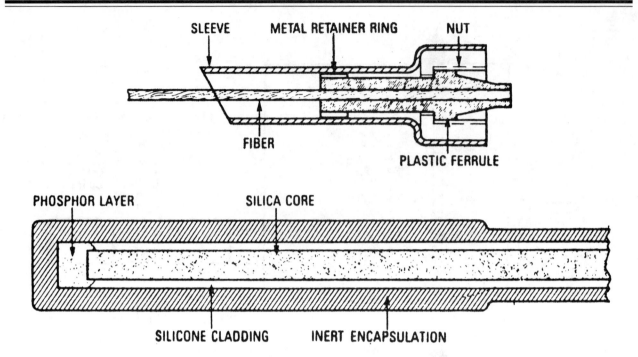

Figure 1-15. *The connector end of the fluoroptic probe (a) and the sensing end of the probe (b).*

mitting temperature measurements in the 50 °C to 200 °C range with an accuracy of +/- 0.1 °C. The additional benefits of this probe are:

1. The probe does not heat in the presence of an RF field.
2. The probe is not electrically or thermally conductive.
3. The long smooth non-contact temperature sensing probe is easily inserted into lab specimens.
4. The probe is chemically inert and can be sterilized.
5. The probe's low thermal mass allows almost instantaneous readings.
6. Calibration is intrinsic to the phosphor probe.

The Fluoroptic Probe

The probe (See *Figure 1-15*) is only 0.7 mm in diameter. But, more importantly, its tip contains a small amount of a rare earth phosphor, europium activated gadolinium oxysulfide. A high intensity ultraviolet (UV) lamp (See L2 in *Figure 1-14*) excites this phosphor by sending this optical energy along the fiber. The light returns to this unique non-contact thermometer via the same optical fiber. But, along the way, a beam splitter creates two separate optical channels. The most interesting is on the right in the diagram. Here, wavelengths of interest are isolated by filters and detected by photosilicon diodes. The signals from the photosilicon diodes are amplified, averaged and converted to a digital signal. A microprocessor calculates the ratio of these two signals and determines the corresponding temperature from a lookup table stored in a read-only memory (ROM). The microprocessor formats this data for the front panel display, the analog output, and the RS-232C output.

There are limitations in the length of this non-contact type optical fiber; however, you can partially overcome them. You can use fibers up to 100 meters if the phosphor were excited by electrons, alpha particles, visible (blue) radiation or radioactive materials contained within the sensor.

Chapter 1 Quiz

1. Ozone depletion can cause skin cancers, T or F.
2. The shorter the UV wavelength, the less biologically destructive, T or F.
3. The sun's UV rays strike the Earth most directly at its poles, T or F.
4. The Earth's natural ozone layer is thickest at the equator, T or F.
5. Ozone destruction most readily occurs on the Earth's surface with the release of chlorine and bromine from synthetic compounds (halo carbons) T or F.
6. Ozone is odorless T or F.
7. The ozone layer acts as a natural UV shield or sunscreen, T or F.
8. Halo carbons can remain in existence for up to 120 years, T or F.
9. A single chlorine atom can destroy up to:
    A. 1,000 ozone molecules.
    B. 10,000 ozone molecules.
    C. 100,000 ozone molecules.
    D. 1,000,000 ozone molecules.
10. UV dose detectors use:
    A. Long integration times.
    B. Short integration times.
    C. Vacuum photodiode detectors.
    D. A and C.
11. One joule per second is defined as:
    A. A Watt.
    B. A Weber.
    C. A Siemen.
    D. An Ampere.
12. You can convert from energy per unit area to power per unit area by equating 1 joule/cm^2 to 1 Watt/cm^2 and:
    A. Squaring the exposure time.
    B. Taking the square root of the exposure time.
    C. Dividing by the exposure time.
    D. Multiplying by the exposure time.
13. The experimental UV index:
    A. Is used by the weather service.
    B. Ranges from 0 to 15 in midsummer.
    C. Is used by the EPA.
    D. All the above.
14. The environment directly above the earth is:
    A. The stratosphere.
    B. The troposphere.
    C. The atmosphere.
    D. None of the above.

15. Which of the following tend to reduce UV intensity?
    A. Clouds.
    B. Rain.
    C. Smog.
    D. All the above.
16. UV affects which of the following animals?
    A. Rodents and birds.
    B. Lizards, iguanas and amphibian eggs.
    C. Sheep.
    D. All of the above.
17. Which of the following are sources of error in a UV dosimeter?
    A. The UV's angle of arrival with respect to the detector.
    B. The temperature of the UV sensor.
    C. A and B.
    D. None of the above.
18. Fire detectors use which of the following optical technologies?
    A. B and C.
    B. Triple IR sensors.
    C. UV and IR.
    D. A pressure sensor.
19. UV based flame and fire detectors are:
    A. B and C only.
    B. Used only inside.
    C. Are subject to false alarms.
    D. Used only outside.
20. Beneficial uses of UV include:
    A. Detecting if a new Benjamin Franklin $100 bill is authentic.
    B. Helping analyze stamps rocks and minerals.
    C. A and B only.
    D. Space guidance applications.
21. The Dobson Unit (DU):
    A. Detects how blue the sky is.
    B. Detects the oxygen content of air.
    C. Detects ozone is a widely distributed area in the sky.
    D. None of the above.
22. Vacuum UV (VUV):
    A. Occurs in the 100 to 200 nm wavelength range.
    B. Is less destructive than UV-C.
    C. Most easily occurs with the aid of helium gas.
    D. None of the above.
23. UV has both positive uses and negative characteristics, T or F?
24. Two negative uses of UV are sterilizing salt water aquariums and a city's waste water, T or F?
25. UV's disinfectant effect results from altering the DNA in a germ, T or F?

Chapter 1

26. Which UV is the most damaging but the atmosphere blocks it?
    A. VUV.
    B. UV-A.
    C. UV-B.
    D. UV-C.
27. Most UV reaches North America in the middle of which season?
    A. Winter.
    B. Spring.
    C. Summer.
    D. Fall.
28. The UV experimental index in summer has what range?
    A. 0 to 15.
    B. 10 to 20.
    C. -5 to 15.
    D. None of the above.
29. In an ideal unpolluted world, there is a constant amount of:
    A. Thermal warming.
    B. Atmospheric heat loss.
    C. Ozone.
    D. Smog.
30. Which animal uses UV to see another of its species' staked-out territory?
    A. Finfish.
    B. Iguanas.
    C. Birds.
    D. Turtles.
31. Which penetrates a substance more than UV curing?
    A. IR curing.
    B. Electron Beam (EB) curing.
    C. Chemical curing.
    D. None of the above.
32. UV based fire sensors analyze a fire by:
    A. Its fuel consumption.
    B. Its oxygen/air composition.
    C. Its evolving heat.
    D. All of the above.
33. The UV Biometer^{tm} uniquely uses what method to determine UV's effect on living organisms?
    A. Bird migration patterns.
    B. The Erythema action curve.
    C. Changes in temperature.
    D. Changes in the air's oxygen content.
34. UV curing dosimeter instruments concern themselves most with the UV curing spectrum of:
    A. The photoinitiator.
    B. The molymers.
    C. The polymers.
    D. None of the above.

35. A flashlamp can produce _____ of Lumens of light in milliseconds.
    A. Hundreds.
    B. Thousands.
    C. Tens of thousands.
    D. Millions.
36. Non-semiconductor flashlamps can:
    A. Detect non-visible corona discharges.
    B. Detect fire and flames.
    C. Both A and B.
    D. None of the above.
37. Normal glass absorbs UV below 300 nm; therefore, you use?
    A. Thicker glass.
    B. Multiple layers of glass.
    C. Thinner glass.
    D. Fused silica or special UV transmitting glass.

Chapter 2
IR Theory & Application: Non-Contact Temperature Measurement

Chapter 2
IR Theory & Application: Non-Contact Temperature Measurement

This chapter introduces IR theory and a representative example of IR in a pyrometer, a non-contact thermometer or radiant heat detector. We begin IR theory with important laws and concepts, as follows:

1. Stefan-Boltzmann's equation.
2. Wien's Displacement Law.
3. Planck's Law.
4. IR Pyrometer principles.
5. The sensor/target distance relationship.
6. Emissivity and blackbody calibration.

We'll also investigate sources of measurement errors and other limitations while surveying two IR measurement case studies; but we'll begin with a brief history of the thermometer and standard accepted temperature scales.

A SHORT HISTORY OF THE THERMOMETER AND ITS SCALES

The First Thermometer

Galileo invented the thermometer circa 1952. This crude device, by today's standards, consisted of a long glass tube with a narrow throat. (See *Figure 2-1*) It housed an open container filled with colored alcohol. As the surrounding air became hotter, the air in the sphere expanded. This caused it to bubble through the liquid. Conversely, as the ambient air cooled, the liquid moved up the tube. This direction of movement opposed mercury within a modern thermometer.

Galileo's invention has therefore been called an "upside down" thermometer. It was a poor indicator of true air temperature since it suffered from both evaporation and changes in barometric pressure; but Galileo's thermometer was nonetheless a clever, innovative, original effort. More importantly, it prompted others to research and improve upon his original idea.

Improved Thermometers

The next improvement in thermometers was the Robert Hook sealed thermometer of 1664, with an ice water point and a healed top. Next, there was the Florentine thermometer. This sealed construc-

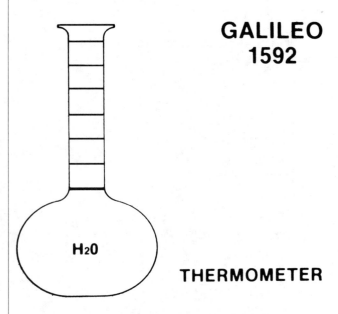

Figure 2-1. The first thermometer, invented by Galileo.
(Courtesy of John Fluke Manufacturing Co.)

Chapter 2

FAHRENHEIT THERMOMETER 1706

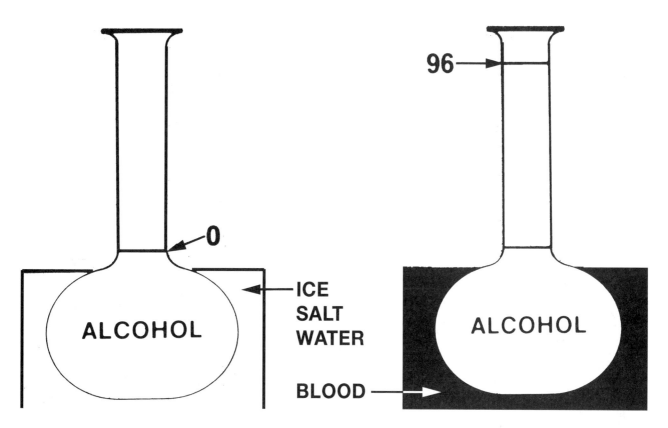

Figure 2-2. The Fahrenheit mercury thermometer. (Courtesy of John Fluke Manufacturing Co.)

tion device had evenly spaced gradients or temperature marks. In the following decades, many thermometers were conceived; but, due to its accuracy, one scale emerged as a standard. This was the Dutch instrument maker Gabriel Fahrenheit's 1706 mercury-based thermometer. (See *Figure 2-2*) This thermometer yielded consistently accurate results. Gabriel Fahrenheit calibrated his thermometer by having its low point immersed in a mixture of ice water and salt (or ammonium chloride). He labeled this as zero degrees and used the temperature of human blood as the high end, assigning it a value of 96. Why 96? Historically, most previous thermometers were marked off in gradients of 12 degrees and 96 is evenly divisible by 12. Another reason may have been that in 1701, the English physicists Newton, who also dabbled with thermometers, said the armpit of a "healthy Englishman" (See *Figure 2-3*) was exactly 12 degrees or marks on his temperature scale. The last major thermometer was the first international temperature scale, in 1887, based on hydrogen gas pressure from temperature expansion and contraction.

Temperature Scales

Around 1742, the Swede Anders Celsius proposed the melting point of ice and the boiling point of water as his thermometer's benchmarks. This created a temperature scale which now bears his name. Interestingly though, initially the melting point was proposed as 100 degrees and the boiling point as zero; however, this convention was later reversed, as we know it today. (See *Figure 2-4*) In 1948, this centigrade "one hundred marks" scale's name was officially changed to Celsius, in honor of Anders Celsius.

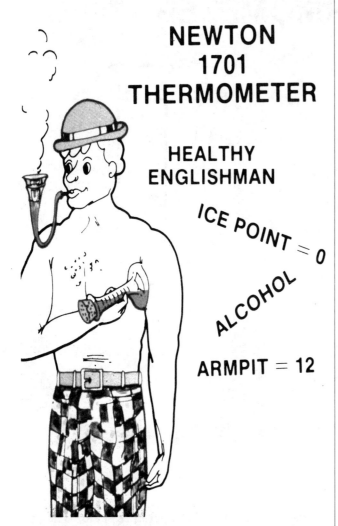

Figure 2-3. Newton's "temperature standard," a healthy Englishman. (Courtesy of John Fluke Manufacturing Co.)

In the early 1800s, William Thomson (Lord Kelvin) developed a universal thermodynamic scale based upon the coefficient of expansion of an ideal gas. Kelvin established the concept of absolute zero, at which all molecular motion ceases (-273.15° C). This scale remains the standard for modern thermometry. (See *Figure 2-5*) The conversion of the following equations are for the four modern temperature scales:

$$°C = 5/9 \, (°F - 32) \quad °F = 9/5 \, °C + 32$$
$$K = °C + 273.15 \quad °R = °F + 459.67$$

The Rankine Scale (°R) is simply the Fahrenheit equivalent of the Kelvin scale and was named after W.J.M. Rankine, an early pioneer in thermodynamics. Note the Kelvin scale does not have a degree sign. It merely is expressed as Kelvins, not degrees.

IR Energy

This is radiation released by an object whose temperature is above absolute zero. That is above -273.15° C or 0 Kelvin. As you heat an object, its molecular activity increases and agitated molecules randomly collide more frequently. This releases detectable heat energy whose quantity the Stefan-Boltzmann equation determines:

$$W = \sigma \, \varepsilon \, T^4$$

W = Energy (total radiation expressed in watts)
σ = Stefan-Boltzmann constant = 5.672×10^{-12} watt-cm^{-2} degrees^{-4}
ε = Emissivity of the target
T = Absolute temperature on the Kelvin scale

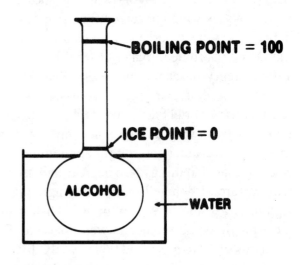

Figure 2-4. The Celsius thermometer. (Courtesy of John Fluke Manufacturing Co.)

Chapter 2

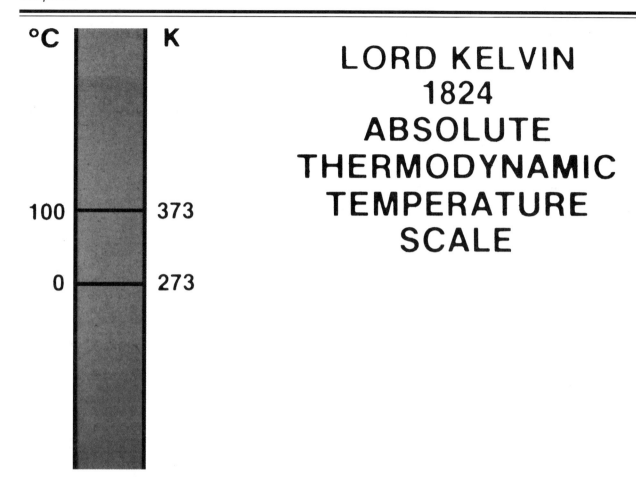

Figure 2-5. The Kelvin thermometer. *(Courtesy of John Fluke Manufacturing Co.)*

Emissivity

Emissivity and absolute temperature influence radiation which is proportional to the fourth power of the absolute temperature. An object at 200° F has 16 times greater radiation than an object at only 100° F. Take two (the ratio of 200° C to 100° C) to the fourth power to derive 16. An IR sensor detects radiation, interprets, and relates it to the object's temperature.

A material's emissivity value also affects the amount of radiation an object emits. Emissivity is the measure of an object's ability to either emit or absorb radiant energy. Emissivity values fall in the range of 0 to 1.0. You empirically determine or obtain these values from tables. (See *Table 2-1*) Emissivity has no unit symbol; it is just a number.

A surface having an emissivity value of "0" indicates a perfect reflector. This type of surface neither emits nor absorbs radiant energy. These well polished, highly reflective surfaces with low emissivity values are not good candidates for IR temperature sensing. Conversely, surfaces which are rough and absorb almost 100% of their radiant energy are good candidates for IR temperature sensing. These surfaces have higher emissivity ratings approaching 1.0. Theoretically, a surface or an object which emits and absorbs 100% of its radiant energy is a blackbody; but since this 100% goal is never quite reached, we sometimes refer to these as *graybodies*.

Blackbody is a slight misnomer. If a surface is black, it does not necessarily have an emissivity of 1.0. Other factors which influence surface emissivity include:

1. Surface texture (its degree of roughness or oxidation).
2. Surface temperature.
3. The wavelength of emitted energy.

When using an IR pyrometer, only point the pyrometer at surfaces with an emissivity value of 0.5 or greater. You can enhance a surface with a low emissivity by the following methods:

1. Texture the surface (i.e. sanding or sandblasting).
2. Oxidize the surface.
3. Anodize the surface.
4. Paint the surface with a dull, highly-absorbent coating.

Electromagnetic Radiation

A surface emits IR energy as electromagnetic waves. These waves are invisible to the human eye, but have the same characteristics as visible light. *Figure 2-6* shows the components of a wavelength. Amplitude is the height of the wave. The distance between two consecutive peaks is the wave's length, or one cycle. The Greek letter λ (lambda) designates a wavelength.

You measure wavelength (λ) in microns, or 0.00004 inches, which is approximately 50 times smaller than a human hair. Wavelength (λ) depends upon the frequency at which molecular activity occurs and the speed of light. The following equation describes this:

$$\lambda = C/f$$

λ is the wavelength, C is the speed of light, and f is the frequency. In IR applications, the higher the frequency, the shorter the wavelength (λ) and the higher the temperature of the sensed object. The IR band has a usable level of intensity for temperature measurement in this 0.5 to 20 micron range.

Non-contact temperature measurement IR sensors intercept radiant energy emitted by an object. This emitted energy directly corresponds to its temperature. These sensors offer numerous advantages over their contact temperature sensor counterparts:

1. You can mount the sensor away from the heat source.
2. You can sense the temperature of a moving object.
3. The sensor will not sink or draw energy (heat) from the sensed object.
4. You can view the object to be measured through a window in a contaminated or explosive environment.

Nongray Bodies

These are objects whose emissivity varies with wavelength and/or temperature. Measuring this type of object can result in false readings. Most IR pyrometers mathematically translate IR energy into temperature. Because an object with an emissivity of 0.7 emits only 70% of the available energy, this would cause the indicated temperature to read lower than the actual temperature. Some IR pyrometer manufacturers attempt to solve this problem with an emissivity compensator. This is a calibrated gain adjustment which increases the amplitude of the detected signal to compensate for energy lost due to an emissivity less than 1. This same adjustment corrects for transmission losses due to viewing through glass, atmospheric conditions, or metal. (See *Figures 2-7a* to *2-7c*) Note the "environment" in *Figure 2-7b*, which may be smog, fog, or a number of partially IR energy absorbing conditions.

As an example, setting the compensator to 0.5 for an object with an emissivity of 0.5 results in a gain increase of 2.0. If you are measuring an object's heat through a piece of glass with a transmission of 40%, the errors are in series. Set the compen-

Chapter 2

Material	Emmissivity
Aluminum (polished)	0.03
Aluminum Oxidized	0.24
Asphalt	0.90
Brick	0.90 - 0.95
Concrete	0.90 - 0.95
Copper	0.04 - 0.30
Glass	0.80 - 0.95
Graphite	0.30
Inconel	0.25
Inconel (heavy oxide)	0.90
Kapton	0.55
Magnesium Oxide	0.95
Mylar	0.90
Paint:	
Krylon Flat Black	0.96
3M Black	0.89
Paper	0.75 - 0.95
Plastic	0.90
Rubber	0.90
Sand	0.75 - 0.90
Silicon	0.35
Teflon	0.45
Tin	0.05
Water	0.93
Wood	0.90 - 0.95

Table 2-1. Typical emissivity values.

sator for $0.5 \cdot 0.4 = 0.2$. The resulting gain is now at a factor of 5. The gain times the compensation factor always equals 1.

Planck's Law

This states you can plot radiated energy as a function of wavelength. The following equation states Planck's Law as:

$$W = C_1 \lambda^{-5}(e^{C_2/\lambda T} - 1)^{-1}$$

W equals watts per square meter per μmeter, C_1 equals $3.74 \cdot 10^8$, C equals $1.44 \cdot 10^4$ and λ is the wavelength expressed in μmeters. T is in Kelvin and *e* is the base of the Napierian or natural logarithm, 2.7183. *Figure 2-8* shows a wavelength to relative energy curve based on the previous equation. It shows the radiation emission curves for blackbodies, graybodies and non-gray bodies. By convention, longer wavelengths are to the right in IR spectral energy charts. This is the reverse from a electromagnetic spectrum chart with which you are probably familiar.

The two simultaneously occurring changes as temperature increases are:

1. The amplitude of the curve increases, increasing the area (energy) under it.
2. The wavelength λ associated with the peak energy (the highest point on the curve) shifts to the shorter wavelength end of the scale.

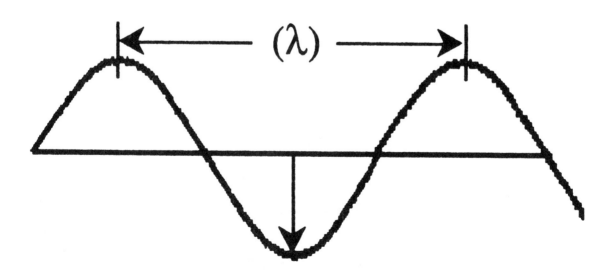

Figure 2-6. A wavelength's components.

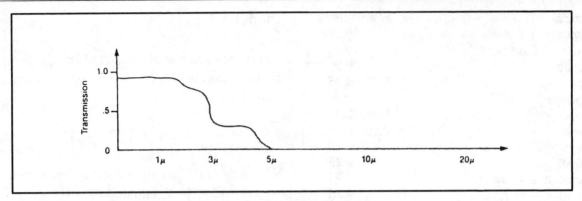

Transmission spectrum of glass (typical 1/8" pane glass).

(a)

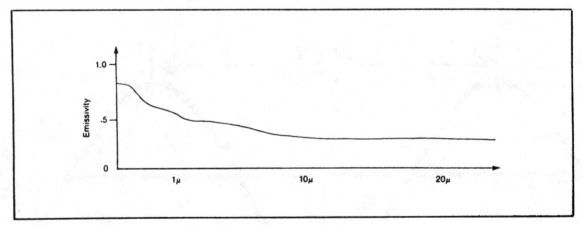

(b)

Typical metal: variation of emissivity with wavelength.

(c)

Figure 2-7. *(a) shows the transmission spectrum of glass; (b) shows radiation as it travels through the atmosphere (Courtesy of Raytek Corp.); and (c) shows the transmission spectrum of metal.*

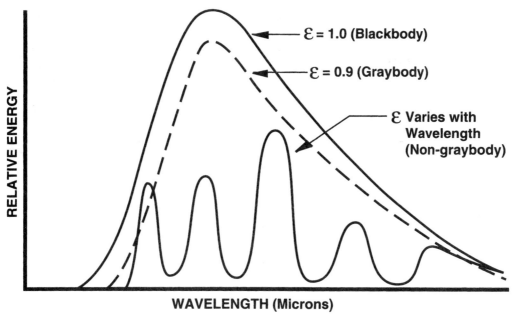

Relative Spectral Distribution Curves of Blackbodies, Graybodies and Non-graybodies

Figure 2-8. *Radiation intensities derived from Planck's Law in the form of blackbody radiation curves.*

Wien's Displacement Law describes this relationship:

$$\lambda^{MAX} = 2.89 \cdot 10^3/T$$

λ^{MAX} = wavelength of peak energy in microns, and T = temperature in Kelvin (not degrees Kelvin).

The following equation describes the wavelength for peak energy emitted from an object at 2617 degrees Celsius (2890 Kelvin):

$$\lambda^{MAX} = 2.89 \cdot 10^3/2890 = 1 \text{ micron}$$

What Happens to Emitted Energy?

Figure 2-9 illustrates what happens to this energy after it has been intercepted by an object. As we showed earlier with the Stefan-Boltzmann equation, the amount of emitted energy depends on both its temperature and emissivity. Once energy is emitted and intercepted by an object, one or more of three things may occur, as described by this equation:

$$E_R + E_A + E_T = 1$$

This is the law of conservation of energy, with E_R = radiated energy, E_T = transmitted energy and E_A = absorbed energy with respect to emissivity. If ε = 0.80, E_A = 80%; then $E_R + E_T$ = 20% for an object with an emissivity of 0.8. Remember, all energies must add up to 1.00.

In a blackbody:

$$e = 1.00, \text{ then } E_R = E_T = 0\%$$

This states all energy in a blackbody is either reflected or transmitted and that no energy is absorbed. This complies with the law of conservation of energy.

In the previous equation, the example states that with an emissivity of 0.8, the object absorbs 80%

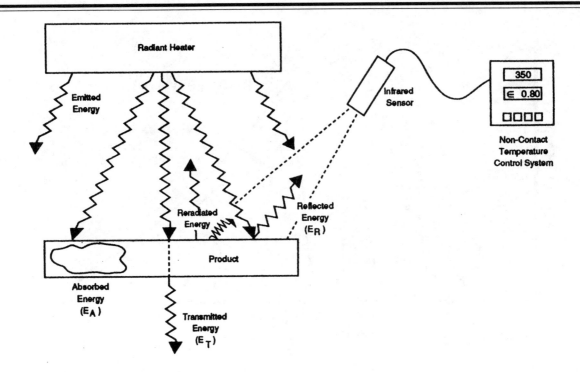

Figure 2-9. Absorbed, radiated and transmitted IR energy. (Courtesy of Raytek Corp.)

of the energy striking it. As an object heats, it starts to re-radiate more energy, and also, its ambient temperature affects accuracy in the reading. *Figure 2-10* illustrates this principle.

How Does An IR Sensor Work?

A non-contact temperature control system consists of:

1. Precision optics.
2. An IR detector.
3. A sensor housing.
4. The support electronics.

As the IR sensor intercepts a portion of the emitted radiant energy, it concentrates this energy into the detector. The detector produces a signal proportional to the magnitude of this emitted IR energy. This signal is amplified, linearized, and conditioned by the support electronics.

Real World IR Pyrometers

The way to categorize many IR pyrometers is by spectral response (the width of the IR spectrum covered). The most common design selects a segment of the IR spectrum and optically filters the pyrometer to only "look" at that narrow sector of the spectrum. You can then integrate the energy falling on the detector for that segment. Many general purpose instruments use a wideband detector, 8 to 14 microns. With adequate energy you require only low gain amplifiers. Some inexpensive units cover the 0.7 to 20 micron spectrum at the trade-off expense of being "distance sensitive" because they include some atmospheric absorption bands. An IR pyrometer operating in the 8 to 14 micron band avoids this problem.

Specialty Pyrometers

For special purposes, very narrow band (2.2 micron) IR pyrometers are available. These are more expensive though because they require more stable,

Chapter 2

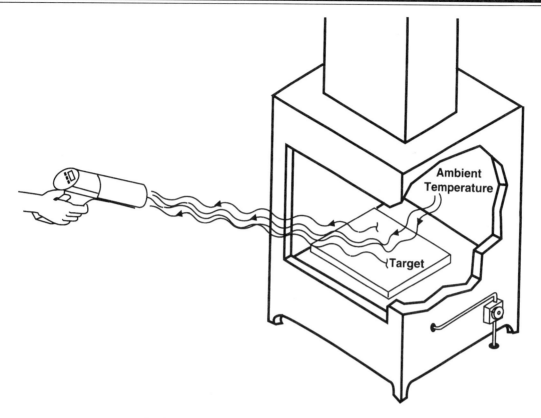

Ambient Energy. High ambient energy in applications such as ovens or furnace walls may result in a temperature reading that is higher than the object. Instruments include a T-ambient function to correct for high ambient energy.

Figure 2-10. *A heated object emitting more radiation due to its high ambient temperature.*

higher gain amplifiers to amplify smaller signals. A third type of IR pyrometer is the two-color pyrometer or ratio type. (See *Figure 2-11*) This instrument measures the ratio of two energies at two narrow bands. If the change in emissivity at the two selected wavelengths is the same, the effect of emissivity is eliminated. An advantage to this type IR pyrometer is that the target need not fill the field of view, which is a requirement of single color pyrometers. If a target with a single color IR pyrometer is cut in half, only half the energy will be received and the instrument indicates a lower temperature. With the two-color IR pyrometer, if both targets' energies are cut in half, the ratio stays the same so no error occurs. This is very handy if, when using a two-color IR pyrometer, a cloud of dust comes between you and the object. The lost energies are no longer a problem since their losses are a fixed ratio; thus the name of this type of IR pyrometer.

Practical Concepts in IR Pyrometer Use

The sensor/target distance is crucial. (See *Figure 2-12*) Position the sensor so the object fills the sensor's entire field of view. In *Figure 2-13*, #1 illustrates proper sensor placement. The sensor is "looking" at the object itself and is not picking up the background radiation. Position #2 illustrates incorrect sensor placement. The sensor is looking at the object as well as the background radiation; therefore, it averages background energy with the energy emitted by the object target. If background radiation is within spectral transmission (8 - 14

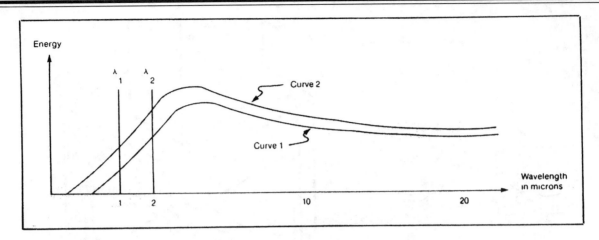

Figure 2-11. Two temperatures, interpreted by the ratio of a two-color IR pyrometer.

microns) of the sensor, an error occurs in the indicated temperature. This background radiation can be from lamps, heaters, motors, heat exchangers, etc. As a general rule of thumb, to minimize the effects of background radiation, the target size should be two times larger than the desired spot size. If an object is 18 inches away and the spot size is rated at 2 inches, the target should be at least 4 inches. *Table 2-2* is a guideline for spectral range measurements.

Sensor Placement

Ideally, a sensor should be placed at a right angle to the target. This helps reduce the effects of reflected energy. Not all applications will lend themselves to perpendicular sensor placement. In these situations, do not position the sensor at more than a 45° angle normal to the surface.

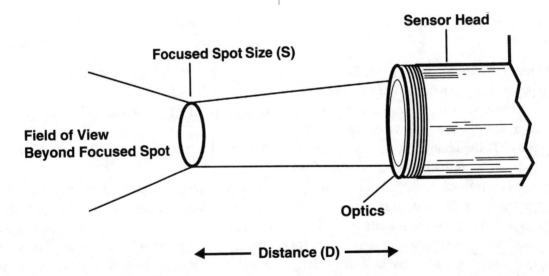

Instrument spot size. Optical resolution of an object is defined as the ratio of the distance from the sensor to the object, compared to the size of spot that is being measured.

Figure 2-12. Target diameter vs. distance from the sensed object. (Courtesy of Raytek Corp.)

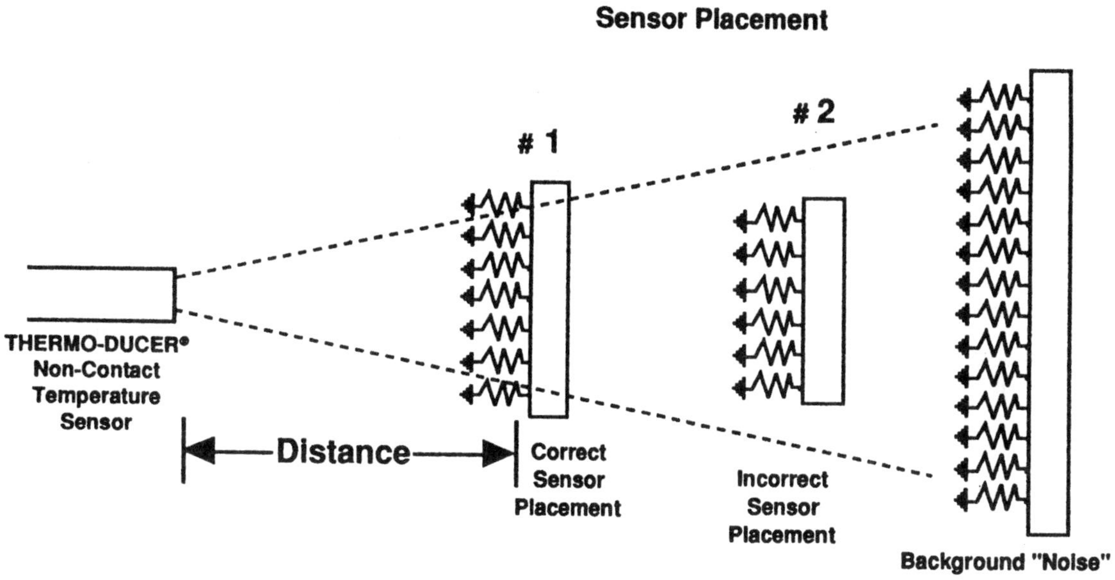

Chapter 2

Sensor Placement

Figure 2-13. Incorrect vs. correct sensor placement from the sensed object.

Reflected Energy and Background Noise

An IR sensor can't distinguish between the object's emitted energy and background radiation. Background "noise" is emitted energy different from the object's temperature. Reflected energy also is a source of error. This energy could be reflected from polished aluminum and/or stainless steel surfaces. Avoid this in your IR temperature measurements. Recall how Chapter 9 stated UV dosimeters purposely used highly polished reflective metals?

Response Times

Thermocouple and RTD contact temperature sensors have response times required to reach 95% or about one time constant (τ). IR technology requires just 240 milliseconds for even detecting a slowly moving object's temperature. You can even reduce the measurement times of faster moving objects when you connect the sensor to additional electronics. The overall system's response is now a function of the sensor's and electronics' time constants.

Sources of Error

Dust, gasses, suspended particles in the air and water vapor all affect IR sensor's performance by absorbing or scattering IR energy. A non-cooled IR sensor indicates temperatures lower than a cooled IR sensor. An IR sensor's output then drifts from exposure to changing ambient temperature. It only stabilizes when the ambient temperature stabilizes.

Viewing Through a Transparent Window

When looking through a transparent window with an IR sensor, you can expect a slight reduction in actual indicated temperature. This is naturally because the glass absorbs a slight amount of energy. However, you can compensate for this by making adjustments when using the emissivity table. (See *Table 2-1*) Realize losses are additive so multiply emissitives.

Calibrating IR Pyrometers

Blackbody calibration is the basis of the accuracy of all IR pyrometers. Blackbodies are used pri-

SPECTRAL RANGE GUIDELINES

1. Choose the shortest wavelength for the temperature range.

2. Use a longer wavelength if a lower temperature is required or the emissivity of the object is higher at the longer wavelength.

3. Be aware of potential problems arising from atmospheric absorption or target transmission.

Table 2-2. Spectral range measuring guidelines.

marily to verify the measurement capability of an IR pyrometer. They, along with other instruments, determine accuracy and performing calibration.

You can use any sources of calibration. However, you must know its temperature and emissivity.

Emissivity is particularly important, especially in the wavelength band intend to measure. Most commercially available blackbodies have an emissivity of 0.97 or greater. The Raytek Blackbody 4000 (See *Figure 2-19* later in the chapter) is a representative blackbody calibrator in the background. The smaller instrument in front of the BB6000 blackbody is a calibrator for such instruments. *Figure 2-14* shows the variables affecting a blackbody calibrator's accuracy.

Surfaces quite commonly change their emissivity over time as a result of the heating they continually experience. Cavity blackbodies, such as the one we are examining, have changes occur in their heating elements or the material itself. Compounding this problem, NIST (formerly the NBS) traceable certification of a blackbody can be made only to its temperature. They can't certify emissivity due to the aging process affecting the blackbody's components. The temperature traceability also stems from the thermocouple or RTD's traceability, not that of the calibrator.

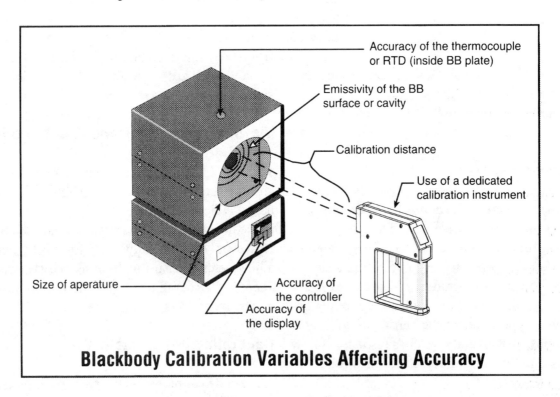

Figure 2-14. Variables affecting a blackbody's accuracy. (Courtesy of Raytek Corp.)

Chapter 2

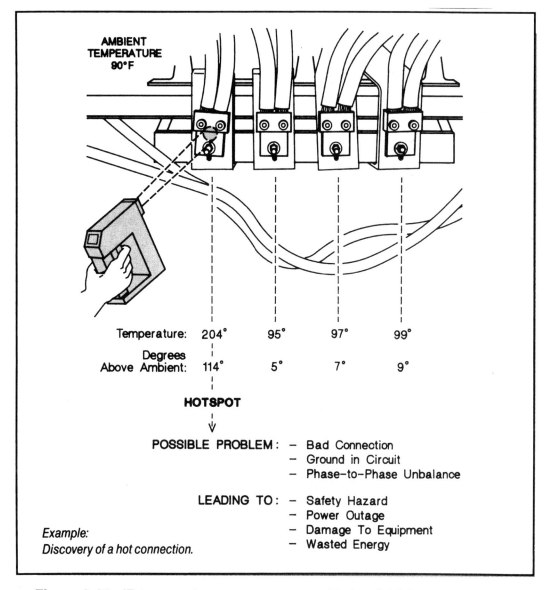

Figure 2-15. *IR temperature measurements of industrial AC power breakers. (Courtesy of Raytek Corp.)*

The accuracy of the blackbody depends upon:

1. Your knowledge of emissivity, which is a changing quantity.
2. The contact measurement device used.
3. The controller and display or meter used — it is necessary to use other means.

This is usually a dedicated calibration instrument.

The Raytek PM3-DC1 instrument checks blackbody accuracy. This is a variation of a standard Raytek PM3. This dedicated instrument uses special high gain optics, a selected detector and comes with a special calibration certificate. It has a temperature measurement accuracy of +/- 0.5%. Note how this PM3 has a visible laser ray produced allowing you to sight or "beam" exactly onto your target. This is the target sight principle used on the guns we sometimes wee in movies!

Case Study #1

Harvard University is the oldest learning institution in American with one building still used which was built in 1720. The routine building mainte-

COMPONENT	POTENTIAL PROBLEMS					SELECTING THE RIGHT TOOL		
	Safety	Load Balance/ Ground in Circuit	Power Dissipation	Winding Insulation Damage	Winding Flaws	Raynger® PM	Raynger® II Plus High Resolution	Raynger® II Plus Long-Range
CONNECTIONS								
Circuit Breakers	✓	✓	✓			●	●	○
Power Panel Terminations	✓	✓	✓			●	●	○
Bus Bars	✓	✓	✓			●	●	○
Fuse Connections	✓	✓	✓			●	●	○
Disconnects	✓	✓	✓			●	●	○
Cable Splices	✓	✓	✓			●	●	○
Knife Switches	✓	✓	✓			●	●	○
Ballasts	✓		✓			◐*	●	○
Battery Bank Terminations	✓		✓			●	●	○
TRANSFORMERS								
Cable Terminations	✓	✓	✓			●**	●	○
Operating Temperature	✓			✓		●**	●	○
Winding Temperature	✓				✓	●**	●	○
ELECTRIC MOTORS								
Cable Terminations	✓	✓	✓			●	●	○
Circuit Breakers/Fuses	✓	✓	✓			●	●	○
Operating Temperature	✓			✓		●	●	○
UTILITIES								
Substation Components	✓	✓	✓			◐**	●	◐
Transformers	✓	✓	✓	✓	✓	◐**	●	◐
Pole-top and Other Distant Components	✓	✓	✓				◐***	●***

● = Excellent   ◐ = Good   ○ = Acceptable   ○ = Not Suitable

* If greater than 10 feet (3.2m) away
** Less than 10 feet (3.2m) away
*** More than 25 feet (8m) away

Table 2-3. Typical potential "hot spots" in electrical wiring. (Courtesy of Raytek Corp.)

nance on some of these buildings is anything but routine. Electricity is purchased at 13.8 kV and distributed at this voltage and lower voltages to individual buildings. The two major utilization voltages are 480Y/277Y for the lighting and motor loads and 208Y/120V for receptacles, lab equipment and incandescent lighting. However, due to their age, the buildings vary greatly and are updated on an "as needed" basis, and no standardization of equipment exists.

Periodic inspections of wiring are made with a pocket-sized IR scanner to localize and identify potential electrical faults. If the lens of an IR thermal scanner has a 50:1 field of view, it can examine a spot 1 inch in diameter from a distance of 50 inches. You can also use a periscope attachment which enables you to scan at a 90° angle. This allows you to reach otherwise inaccessible areas, such as the interior of panel boards and hidden parts of switchboards.

Figure 2-15 illustrates measuring the temperature of AC breakers with the wall panel removed. The test began in a kitchen at 7 AM because that was the time for full load use. All circuit breakers were scanned twice and all proved satisfactory except #25, which indicated in the red zone. This circuit breaker was a single-pole 20A, 120 V unit wired with a #12 AWG THHN conductor. A clamp-on ammeter indicated about a 28 A load. Further investigation showed that loads had been gradually added to this unit, but it had never tripped.

The scanner further indicated that the breaker was hot and that the heat continued at the wire termination and along the conductor. More localized heating would have indicated a defect in the breaker. The breaker, though, was replaced and the loads on this one unit were redistributed in a more equitable fashion. *Table 2-3* summarizes the typical faults to inspect in industrial AC wiring.

Chapter 2

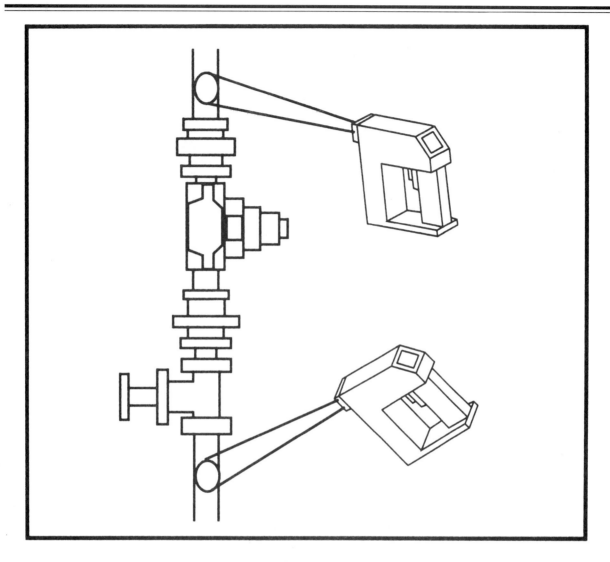

Figure 2-16. *Measuring the temperature of a steam trap line's inlet and outlet.*

Case Study #2

Industrial steam distribution systems require periodic trap maintenance. If not properly maintained, they can eventually blow. If not properly maintained, they can eventually blow. A blown trap passes steam at a rate directly proportional to the inlet pressure and orifice size. Diagnosing a steam trap's problems by merely watching it or by removing the condensate return line is labor intensive. An IR pyrometer though can measure the temperatures on the inlet and outlet steam lines. (See *Figure 2-16*)

Proper trap operation has a high temperature inlet and a low temperature outlet. The inlet temperature is directly proportional to saturated steam pressure. (See *Table 2-4*) If the valve is blown, you'll have both high temperatures on the inlet and outlet since steam is just blowing through. If a valve is plugged, you'll have both low input and outlet temperatures since the condensate return line has filled the trap and is filling the inlet line. *Table 2-5* summarizes trap line faults and symptoms.

Other Trap Problems

Since factories tend to grow, and not always as planned, you may experience added demands on

Table 1 - Relationship between pressure and temperature					
Absolute Pressure		Gage Pressure		Temperature	
osi	KPa	(*) osi	KPa	(*)F	C
14.7	101	0	0	212	100
20	136	5	34	228	109
25	170	10	69	240	116
30	205	15	103	250	121
35	239	20	138	259	126
40	274	25	172	267	131
45	308	30	207	274	134
50	343	35	241	281	138
55	377	40	276	287	142
60	412	45	310	293	145
65	446	50	345	298	148
70	481	55	379	303	151
75	515	60	414	308	153
80	550	65	448	312	156
85	584	70	483	316	158
90	618	75	517	320	160
95	653	80	552	324	162
100	687	85	586	328	164
110	756	95	655	335	168
120	825	105	724	341	172
130	894	115	793	347	175
140	963	125	862	353	178
150	1032	135	931	358	181
160	1101	145	1000	364	184
170	1170	155	1069	368	187
180	1239	165	1138	373	189
190	1308	175	1207	378	192
200	1377	185	1276	382	194
250	1722	235	1620	400	214
300	2066	285	1965	417	214
350	2411	335	2310	432	222
400	2756	385	2655	445	229
450	3101	435	2999	456	326
500	3445	485	3344	467	242

(*)Calculations for other quantities in table are based on values from starred columns.

Table 2-4. Saturated steam variables.

steam distribution systems. Lines are often overloaded and extended beyond their capacity. The following is a four-step corrective action procedure for this:

1. Replace faulty insulation or insulation where needed.
2. Install new steam supply lines to meet overloaded capacity.
3. Re-evaluate present use of steam on overloaded lines to see if they are absolutely necessary.
4. Check operation of the steam supply and condensate return lines to locate problem areas due to defective valves, regulators, heat exchangers, traps and coils. Return lines may have their valves mistakenly shut, causing a "deadhead" condition.

Not All Traps Are Alike

Some traps vary in their construction and properties. Let's take a quick overview of the six most common types. An inverted bucket trap is normally filled with condensate. This maintains the seal around the **inverted bucket**. When live steam enters the trap, the bucket floats and closes the valve. The **float and thermostatic** type trap has condensate enter the trap. This causes a ball float to rise and position the modulating valve where it

Diagnostic Summary Chart

Trap Condition	T Inlet	T Outlet
Normal	Hot	Cold
Blown (Open)	Hot	Hot
Plugged (Shut)	Cold	Cold

Table 2-5. Trap line faults and symptoms. (Courtesy of Raytek Corp.)

passes condensate continuously as it enters the trap. These type valves have a high thermal coefficient since they never release steam.

A **bi-metal** trap has a thermostatic element which allows discharge of air and condensate until the condensate reaches a preset temperature. The **balanced pressure** thermostatic trap operates by a thermostatic fluid filled bellows. This contracts or expands, depending on the trap temperature. **Liquid expansion** thermostatic traps, upon startup, have their air and condensate discharged until the condensate reaches a preset temperature below the boiling point of water (100° C). The liquid-filled thermostatic element then throttles the valve to maintain the preset condensate discharge temperature. Lastly, the **thermo-dynamic** trap starts up with air and condensate passing through the trap. When steam reaches this type trap, velocity under the trap increases and the disk snaps shut. As the steam/condensate cools, the chamber temperature decreases and the disk snaps open to pass condensate.

Diesel Pyrometer Troubleshooting Examples

The temperature differential technique is measuring and then comparing two different spots. Newer pyrometers have a "Use Stored Data" setting with a "Yes" button. Point one of these newer IR pyrometers at the first measurement area and pull the trigger for several seconds. Aim at the second target and perform this same procedure. The difference is in the first and second readings of the inlet and outlet to a diesel's radiator. (See *Figure 2-17*) *Figure 2-18* shows a diesel's oil cooler. Mechanics use these temperature drops in determining a diesel's proper operation.

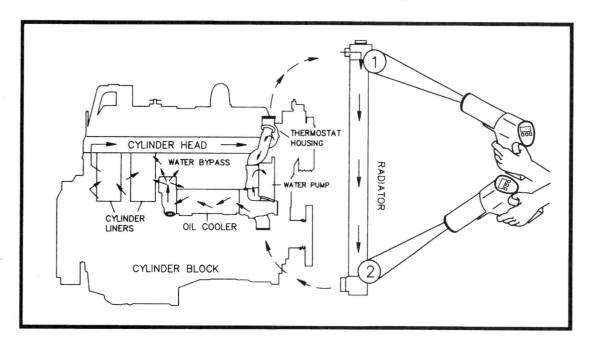

Figure 2-17. Measuring a diesel radiator's inlet and outlet.

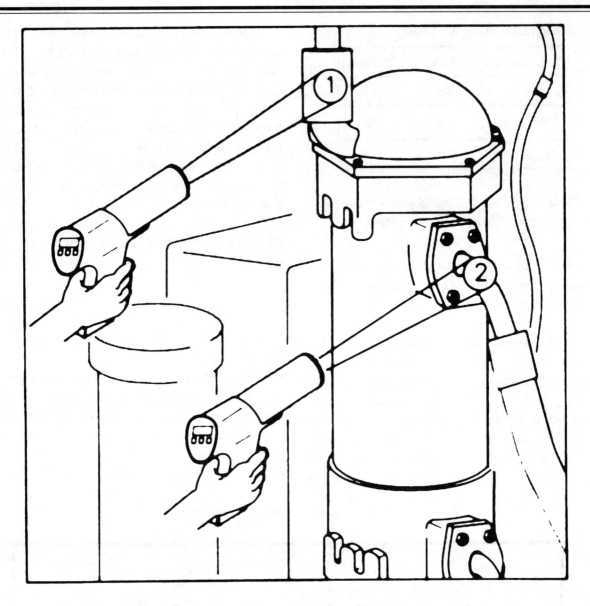

Figure 2-18. Measuring a diesel's oil cooler.

Actual Pyrometer Instruments

IR temperature sensors span a wide variety of instruments. IR pyrometers vary in size and options with some IR pyrometers with laser sights and eyecups to reduce glare from high ambient source of light. These include the Raytek Thermalert 2C series. (See *Figure 2-19*) These ratio pyrometers come in three models which have from 44:1 to 130:1 optical resolution. There are also chart recorders, and control boxes to set trip points, intervals of sampling and alarms in various forms. *Figure 2-20* shows this relationship/interaction. The chart recorders graph temperature and set points over the time you specify. The instrument records, stores and recalls complete time/temperature graphs.

Pyrometer Software

Some representative software is the Raytek Field Calibration and Diagnostic Software™. This software is for the Thermalert™ 2C series pyrometers and allows you to:

1. Recalibrate sensors on-site for a specific target or process temperature.

Chapter 2

Figure 2-19. The Thermalert™ 2C ratio pyrometer in the foreground with the BB1000 black body calibrator in the background. (Courtesy of Raytek Corp.)

2. Perform hardware autodiagnostics to eliminate hardware as a source of error in troubleshooting.
3. Select one, two or three-point calibration, or you can use the factory default settings.
4. Print an audit trail of calibration histories.
5. Support remote sensors and on-screen monitoring through the RS485 interface.
6. Scale the 4-20 mA output.

EXAMPLES OF INFRARED DETECTORS

Type	Examples
Photoconductive	•Mercury-Cadmium-Telluride
Photovoltaic	•Silicon •Indium-Gallium-Arsenide
Pyroelectric	•Lithium-Tantilate
Thermovoltaic	•Thermopile

Table 2-6. Examples of IR detectors.

IR Sensor ICs

Table 2-6 is a summary of IR detector technologies. However, if you want to experiment with IR, you can build a simple human or animal intrusion alarm based on the Hamamatsu P2613 or P2288 (dual detector) series of ICs. See *Figure 2-21* for the IC's basic interconnection. When IR hits their open surface, these ICs experience an IR induced charge and feature a built-in high impedance FET (Field Effect Transistor). Keep the source resistance in the range of 10 kW to 50 kW

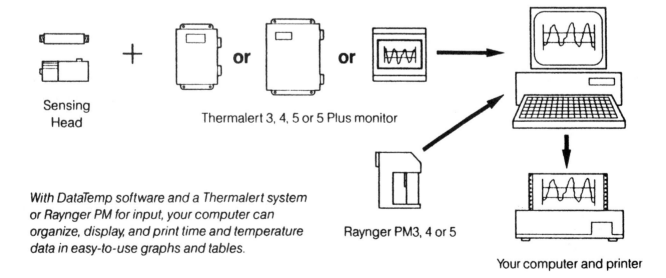

With DataTemp software and a Thermalert system or Raynger PM for input, your computer can organize, display, and print time and temperature data in easy-to-use graphs and tables.

Figure 2-20. The relationship between the IR pyrometer's thermal sensing head, the monitor and a PC. (Courtesy of Raytek Corp.)

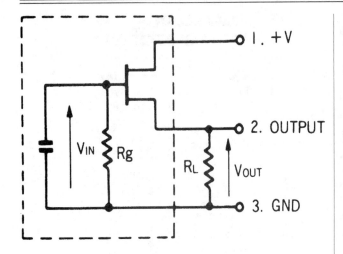

Figure 2-21. An IR sensor IC and its basic connections. (Courtesy of Hamamatsu)

for best results. The human alarm IR detector (See *Figure 2-22*) detects the body's IR, and you amplify this signal and bandpass-filter it to approximate the angular velocity appropriate for humans. The circuit in *Figure 2-22* has a passband of 0.3 to 7 Hz and an amplifier gain of 145 at 1 Hz. By cascading two stages, you realize an approximate gain of 20,000. To complete this circuit, add the alarm and driver of your choice. *Figure 2-22* includes the amplifier gain and cutoff frequency equations.

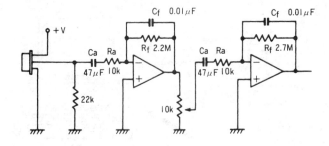

Low-cut frequency $f_L = \dfrac{1}{2\pi C_a R_a}$

High-cut frequency $f_H = \dfrac{1}{2\pi C_f R_f}$

Figure 2-22. An IR IC based intrusion alarm to detect humans. (Courtesy of Hamamatsu)

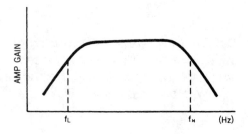

Amplifier gain $A = 1 + \dfrac{\omega C_a R_f}{(1 + \omega C_a R_a)(1 + \omega C_f R_f)}$

Chapter 2

Chapter 2 Quiz

1. Who invented the first thermometer?
   A. Kelvin.
   B. Galileo.
   C. Fahrenheit.
   D. Celsius.
2. Who followed Galileo's thermometer with the most advanced instrument for its time?
   A. Kelvin.
   B. Galileo.
   C. Fahrenheit.
   D. Celsius.
3. Who proposed a temperature scale starting at absolute zero?
   A. Kelvin.
   B. Galileo.
   C. Fahrenheit.
   D. Celsius.
4. Who first proposed a temperature scale with zero degrees as the boiling point of water and 100 degrees as its freezing point?
   A. Kelvin.
   B. Galileo.
   C. Fahrenheit.
   D. Celsius.
5. One whole cycle of a wavelength is its:
   A. Period or wavelength.
   B. Frequency.
   C. Both A and B.
   D. None of the above.
6. A surface with an emissivity of _____ is a perfect reflector of IR energy.
   A. 0.50.
   B. 1.00.
   C. 0.
   D. 0.1.
7. Radiated, absorbed and transmitted energy = :
   A. 0.5.
   B. 1.00.
   C. 0.
   D. None of the above.
8. A two color pyrometer is also called what?
   A. A fixed focus pyrometer.
   B. A ratio pyrometer.
   C. An adjustable pyrometer.
   D. None of the above.

9. Making the target size twice as large as the desired spot size ensures accurate readings by:
    A. Minimizing the effects of background radiation.
    B. Maximizing the effects of background radiation.
    C. Both A and B under different conditions.
    D. None of the above.
10. IR temperature measurements should not be made from shinny stainless steel surfaces, T or F?
11. A cooled IR sensor indicates temperatures less than a non-cooled IR sensor, T or F?
12. NIST certifies a blackbody's emissivity, T or F?
13. Proper steam traps should have a low input temperature and a high output temperature, T or F?
14. A bi-metallic trap has a thermostat which discharges air and condensate until it reaches a pre-set temperature, T or F?
15. An IR sensor based human intrusion alarm must bandpass filter its input signal to approximate the angular velocity appropriate for humans, T or F?

Chapter 3
IrDA and Bar Code IR Applications and Wrist Instruments

Chapter 3
IrDA and Bar Code IR Applications and Wrist Instruments

This chapter investigates two IR applications:

1. The 1996 IrDA standard for PC-to-peripherals (mostly printers) IR communications.
2. Bar graph scanning.

It also investigates both optical and non-optical wrist instruments, some of which are IR based.

Optical Communications and the IrDA Standard

PDAs (Personal Digital Assistants) like the Omnibook, Sharp Wizard™, HP 95LX, Zoomer, or the Apple Newton™ have manufacturer specific non-interchangeable, incompatible IR peripherals. Serial IR (SIR) in computers has been available since 1993, but totally lacks standardization. *Figure 3-1* is an H-P calculator and serial IR printer, circa 1990. It used their own IR standard. The 1996 IrDA (**I**nf**r**ared **D**ata **A**ssociation) standard addressed this.

It seems so simple; pulse modulate an IR LED transmitter with a wavelength matched IR receiver on the other end. However, increasing transmission speed, sophisticated protocols, and the light-to-electrical transformation can garble transmissions. One of IR's benefits is a faster, less expensive ($4 compared to a $20 RS-232C) interface technology. *Table 3-1* partially lists the rapidly growing IrDA members, which parallels the IrDA's growth in speed. In 1993 it was 115 kbits/sec, in 1994 it was 1.2 Mbits/sec and in 1996 it reached 4 Mbits/sec.

IrDA's Competition

RF and IR wireless technologies only slightly compete. Unlike RF, IR communications don't require FCC approval and concentrates on connectivity, not communications. Unlike an LAN's high powered diffuse network goal, IrDA seeks low power, line-of-sight, unobstructed short-range connectivity and a narrow field (30° transmitter to receiver angle).

Figure 3-1. A Hewlett Packard calculator, IR interface module and IR compatible printer. (Courtesy of Hewlett Packard)

Actisys	Genoa Technology	Parallax Research
Adaptec	Geoworks	Philips
Adv'd Micro Devices	GES Singapore	Photonics
Alps Electronics	Hewlett Packard	Plantronics
Alroma Scientific	Hitachi	Polaroid
AMP	IBM	Protell Comm
Apple Computer	ICI Personal Systems	Psion
AST Research	II Standley	Puma Technology
AT&T - AT&T/GIS	Intel	Questor Software
British Telecom	Inventec	ROHM
Brother Int'l	Irvine Sensors	Samsung
Canon	ITT Cannon	Sanyon
Casio Computer	K&M Electronics	Scientific Techn
Cirrus Logic/Crystal	Lexmark Int'l	Seiko
Citizen	Logictech	Sharp Electronics
Compaq Computer	LXE	Siemens
Compression Labs	Marquette Electronics	Sony
Connexus	Matsushita/Panasonic	Standard Micro Corp
DaeWoo Telecom	Megasoft	Stenograph
Dell Computer	Microsoft	Sumitomo Electric
Digital Equipm't Corp	Mitsubishi	Sun Microsystems
DOWA	Motorola	SystemSoft Corp
Eastman Kodak	MPR Tecltech	TDK/Silicon Systems
ECI France	National Semi	TeleQual
Ericsson	NEC	TEMIC/Daimier Benz
Executone	Nokia Mobile Phones	Texas Instruments
Extended Systems	Norand	Timex
Farallon Computing	Northern Telecom	Toshiba
Farpoint Comm	Novell	Traveling Software
Forte Comm	NTT/Nippon Tel&Tel	Unitrode
Fujitsu	Okaya Systemware	VLSI Technology
Gateway 2000	OKI Electric	Windbond
Gemplus	Olivetti	Xerox
General Instruments	O'Neil Software	
General Magic	Open Interface	

Table 3-1. A partial listing of the growing IrDA members.

Implementing the IrDA

IR remote TV controls began in 1974 in Germany. Today, IR technology controls VCRs, audio equipment, automobile theft, keyless entry devices. Two Telefunken ICs provide a rare example of IR and RF competing in these car security systems. *Figure 3-2* is the U2740B, an RF PLL (Phase-Locked Loop) 10 to 100 meter 200, to 500 MHz AM or FM transmitter. This single 16-pin SOIC is for keyless car entry and other short distance data communications such as wireless mice and keyboards. The IC has a microprocessor compatible interface and requires a single time base crystal for the RF oscillator. This approach is less expensive but outperforms SAW (Surface Acoustic Filter) systems. *Figure 3-3* is the U2270B, a single IC reader for car immobilizer antitheft systems. It's reader coil driver circuits are capable of driving 200 mA but consume a scant 40 µA in the standby mode. Typically, it works in concert with a Telefunken e5530PC or e5530GT transponder, a device which reacts when interrogated with the proper code and/or frequency, and Telefunken's M44C260 microcontroller. (See *Figure 3-3*) The U2270B emits a 175 kHz RF field which powers the transponder tag embedded in the driver's car key. The transponder requires no other power source. These

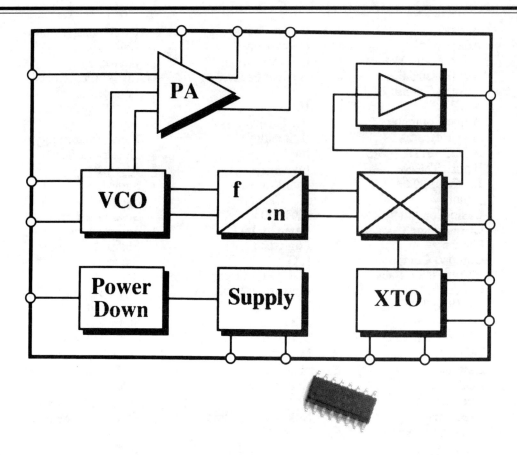

The U 2740 B single-chip PLL transmitter provides a compact solution for automotive keyless entry and other small form-factor short-distance data communication applications. (*Photo courtesy of Telefunken Semiconductors.*)

Figure 3-2. A U2740B single IC PLL (Phase Locked Loop) automotive keyless entry system using RF data communications. (Courtesy of Telefunken Semiconductors)

new ICs only compete within the car security market, not in the IrDA's main realm of connectivity. This is not to imply that Termic Telefunken does not make IrDA ICs. Their TFDS6000 IR transceiver is a compact highly integrated IC designed to implement IrDA links.

IR Still Competes

Typically, a single IC with a highly efficient emitter of ultra-pure III-IV compounds contains a complete IR connectivity system of a photo PIN diode and an amplifier. The silicon detector PIN diodes have very low capacitance and high dynamic sensitivity at long IR wavelengths.

The previous IrDA standard serially transmitted up to 115 kbits/sec over 3 meters. The new IrDA transmits up to 4 MBits/sec. Transfers at 1 Mbit/sec or above require a 880 nm IR; however, most can still converse with older 950 nm devices. The IrDA format transmitter limits duty cycle on the light pulse via the shift register and flip-flop interaction. GaAlAs IR LEDs transform data into light energy (photons).

Their digital output represents the direct image of the electrical input to the transmitter. Only the active low bits (0) are transmitted. On the other end, a digital, pulse-shaping circuit shortens the pulse to 3/16 the emitted bit length.

Chapter 3

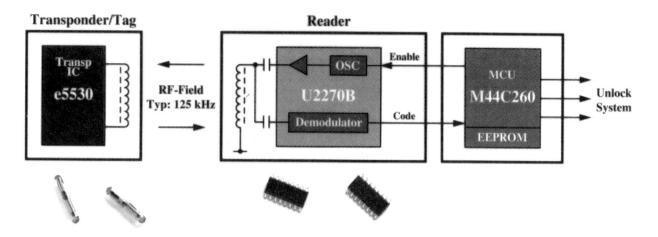

Figure 3-3. A U2270B single IC for automobile immobilizer anti-theft systems.
(Courtesy of Telefunken Semiconductors)

The receiving photodiode and lens produce photons which are converted to current, digitized and amplified. The new standard's 4 Mbits/sec over 3 meters will likely use just 1 meter to save battery life. PDAs run on AA or AAA batteries and have a $5 serial IC price limit.

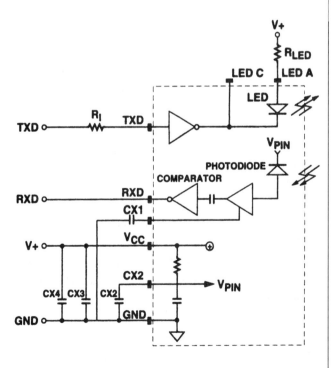

Figure 3-4. A typical configuration of an IrDA transceiver.

Three Representative IrDA ICs

These examples are:

1. The H-P HSDL-1000.
2. The Linear Technology LT1319.
3. The Novalog SIRFIR.

The H-P IC is for subnotebook computers and peripheral devices. The H-P HSDL-1000 serial IrDA compliant transceivers consume just 1.1 mA with a logic high. They directly interface with selected I/O ICs performing PWM (Pulse Width Modulation) and demodulation. *Figure 3-4* is a typical IrDA configuration. The IrDA IC is within dotted lines.

The IrDA Physical Layer Specification defines reliability of connectivity as having a minimum receiver sensitivity of 4.0 mW/cm^2 at a bit error rate of 10^{-9} in the presence of:

1. 10 kilolux of sunlight.
2. 1,000 lux of fluorescent or incandescent light.

Maximum light reaching the detector IR LED depends on its housing's shape.

Linear Technology's LT1319 IC is for multiple modulation, standard IR receivers and transforms

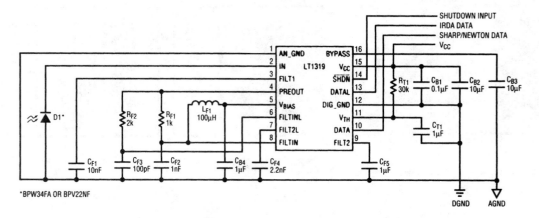

Figure 3-5. *A Linear Technology LT1319 IrDA transceiver and Sharp/Newton data receiver. (Courtesy of Linear Technology)*

modulated photodiode signals back to digital signals. *Figure 3-5* shows an IrDA and a Sharp/Newton PDA data receiver. *Figure 3-6* is a detailed block diagram of this low noise, high speed, high gain IR receiver IC. It takes the IR photodiode current from either a Siemens BPW34FA or Temic BPV22NF for current-to-voltage conversion. After external filtering, depending on the communication standard you use, there are two additional filter buffers. The thresholds of the comparators are set externally by the current flowing into the VTH pin. The high frequency comparator has a response time of 25 ns for high data rates.

The low frequency comparator responds in 60 ns for compatibility with the Sharp/Newton PDA and IrDA standards. Its internal shutdown circuitry saves power. AC coupling loops around the preamp and the two gain stages help reject ambient interference. An external RC pair sets the rejection frequency. The internal supply regulator helps reject power supply noise.

Extending the Transmission Range

You can extend this IC's range by compensating for the peak current at the output of the preamp and the input at pin 2, which will sag. Add an NPN transistor with its emitter tied to pin 2, and its base to pin 4, and its collector to the +5 volt supply.

(See *Figure 3-7*) A Sharp or Newton PDA IrDA interface uses an ordinary 2N3904 with a bandwidth of 2 MHz. If you want to use the faster SIR interface you will need a faster transistor.

LED Drive Circuits

There are several simple circuits for driving IR LEDs. A 2N3904 transistor in a SOT-23 package is fine for the Sharp and Newton PDAs' lower modulation speeds with pulses over 1 msec. (See *Figure 3-7*) The 16 ohm resistor limits instantaneous peak currents to 200 mA. Use a 10 µF low ESR bypass capacitor and a 0.1 µF capacitor to bypass the connection to the LED so its supply voltage does not temporarily sag under a high current pulse.

Pulse widths less than 500 nsec require an N-channel MOSFET with an on-resistance of less than an ohm when operated with 5 volts on the gate. The FET turns off much quicker than the saturated NPN bipolar transistor, providing less ON-resistance. *Figure 3-8* is a circuit layout which includes two devices available in the small SPT-23 packages. Liberal copper on both sides of the PC board accommodates high speed switching.

The Irvine Sensors SIR2 (See *Figure 3-9*) consumes a scant 90 µA current from either 3.3 or 5

Chapter 3

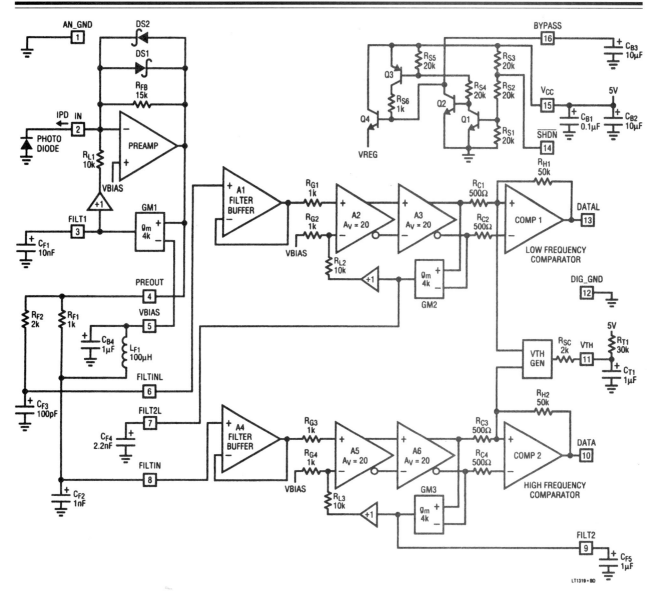

Figure 3-6. A detailed block diagram of the Linear Technology LT1319 IrDA transceiver. *(Courtesy of Linear Technology)*

volts. This IC interfaces wide dynamic range IrDA compatible detector IR photodiodes and drives 40 pF at CMOS/TTL levels. This permits IrDA compliance with UARTs or National Semiconductor's Super I/O devices from 2,400 to 115.2 kbaud. *Figure 3-10* shows how a Novalog SIRIF IC connects to a Super I/O device or UART.

Making Your Own IR Filter

To experiment with IrDA ICs, consider making your own IR optical filter. Pull the film completely out of a color roll of ASA 100 film, purposely exposing it to sunlight for several seconds. Roll the film back into the roll. Process the exposed film, making a note for the processor not to print frames. When double layered, the processed film is an excellent optical filter for blocking ambient light and passing IR.

Before a manufacturer can use the IrDA beaming logo (See *Figure 3-11*) they have to provide a signed statement that they have thoroughly tested their IrDA device. This is testing of both software and hardware.

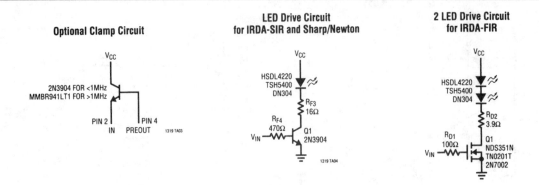

Figure 3-7. Extending transmission range by various external transistor drive schemes.

SERIAL IR HARDWARE AND SOFTWARE

IrDA Software

Let's examine three representative products from different companies which ease SIR IrDA testing. TranXit Pro™, from Puma Technologies, is software for printing and transfer of IrDA based files. The TranXit software, available in 16 Western and Asian languages, has five salient features:

1. You can create and save synchronization "profiles." A profile contains the directories to be synchronized and the options to be used, and can be executed automatically whenever specific PCs are connected.
2. Wireless printing is possible with popular printers such as the HP LaserJet™ 5P, or you can use an IR connector (to be covered shortly) and TranXit's Print Redirector for access to the desktop PC's printer.
3. Wireless Clipboard transfers files between Windows™ applications with the usual cut-and-paste commands.
4. Data File Transfer greatly improves performance by only sending changes made to a file, not the entire file.
5. The "Traffic Light" icon allows real-time monitoring of your serial IrDA's connection status so you won't miss any data transfers.

Figure 3-12 is a block diagram of this software which runs on a variety of operating systems such as Windows™ 3.1, Windows™ 95, Windows™ for Work Groups, Windows™ NT, OS/2 Warp, and Mac OS. *Figure 3-13* shows a greatly simplified example of how an IrDA protocol negotiates a baud rate and establishes a connection. Note the initial transfer took place at 9.6 kbits/sec.

PC Board Layout for IRDA-SIR/FIR and Sharp or TV Remote Data Receiver

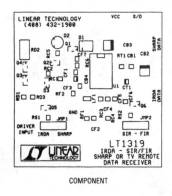
COMPONENT

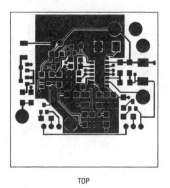
TOP

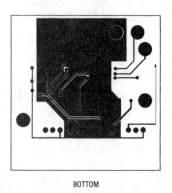
BOTTOM

Figure 3-8. A PC board layout of a Linear Technology LT1319 IrDA transceiver and Sharp/Newton data receiver. (Courtesy of Linear Technology)

Serial IR Example Schematics

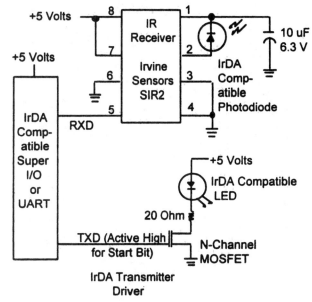

Figure 3-9. The Irvine Sensors SIR2 IrDA IC.

Development IrDA Software

If you develop IrDA software you'll need protocol conformance testing. Genoa Technology has five software programs produced by Genoa technology which help ensure that your hardware and/or software complies with the IrDA standard.

The IrLAP Test Suite consists of 385 test cases for minimal secondary (responding) implementation. Testing of the information exchange procedure alone consists of 250 test cases. The test cases methodically cover the state transitions and variables defined in the IrDA protocol specifications. Since a secondary device must also support the IrDA protocol specifications and the information exchange, this software covers most of the primary (commanding) functions as well.

The second test software is IrLMP Test Suite which consists of 194 test cases covering link management connection and data transfer, as well as the

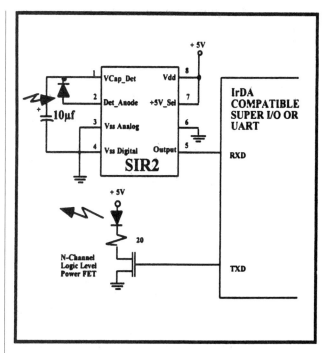

Figure 3-10. A Novalog SIRIF IrDA IC connected to a Super I/O or UART. *(Courtesy of Novalog)*

Figure 3-11. The IrDA compliance logo.

Information Access Service. The Test Report software uses a monitor queue to capture every inbound and outbound frame transmitted during the execution of a test case. When the test is complete, the queue is analyzed for timing violations.

The Bit Error Rate (BER) Tester software provides the tools to test the reliability of the physical layer at various transmission angles, distances, power levels, and lighting conditions. The BER tester uses echoing of test frames to provide the data flow in both directions. You have complete control over the data frame size, pattern, and the duration of the test.

ACTiSYS Corp. makes four different test software packages. The first is IR900SW, a special IrDA frame tester which also lists error rates at different speeds, distances, and angles. The IR920SW is an IrDA protocol software package for testing printers, cellular phones, PBX, etc. and is in the 80C51 microcontroller code. The IR940SW is also an IrDA protocol software package for testing printers, cellular phones, PBX, etc., and is in C code.

The IR960SW is a protocol software package for testing palmtop computers running on DOS or RISC (Reduced Instruction Set Controllers). Its C code can be embedded into an ASIC IC.

IrDA Based Hardware

There are interfaces which convert non-IrDA based products to this standard. The ACTiSYS ACT-IR3D (See *Figure 3-14*) transfers data via an IR link and prints data from your IR capable PC or PDA. It communicates from speeds of 9,600 to 115,200 bits per second. It can detect and select the IrDA standard or Sharp's ASK-mode modulation protocol or speeds. The device has serial DB-9F computer and parallel DB-25F connectors on each side.

The ACT-IR3S+ (See *Figure 3-15*) has in the lower portion of the photo, from left to right, a Sharp Wizard/Zaurus, an Apple Newton, and an IrDA notebook PC all transmitting (not simultaneously) to the ACT-IR3+. This allows you to have an IR computer and printer link in one. This device transfers files to desktop PCs or MACs for record

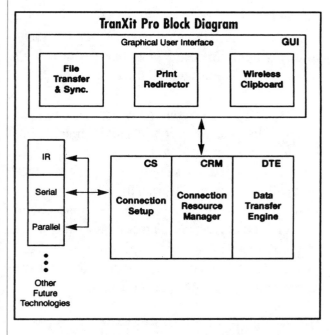

Figure 3-12. A block diagram of Puma Technologies' TranXit Pro^{tm} software. (Courtesy of Puma Technologies)

Simplified Example of IrDA Protocol "Negotiating" a Baud Rate and Establishing a Connection

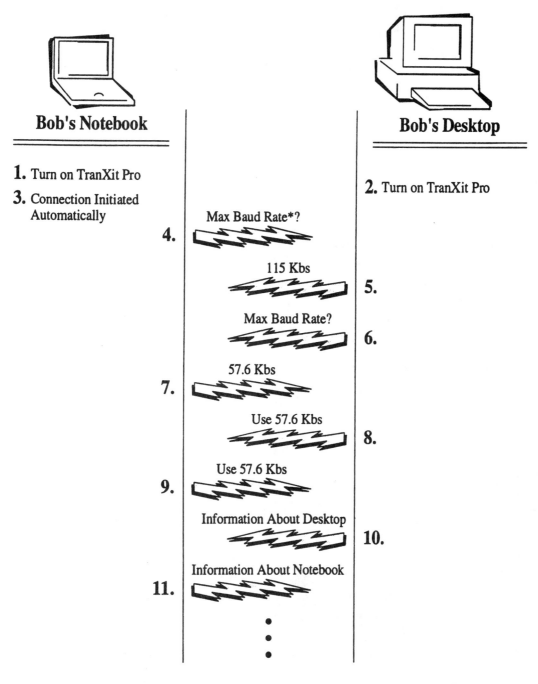

Initial transfers take place at 9.6 Kbs

Figure 3-13. A simplified example of how IrDA negotiates a baud rate to establish connection.

Figure 3-14. The ACTiSYS ACT-IR3D IR data linker for IrDA compatible PC or PDA. (Courtesy of ACTiSYS)

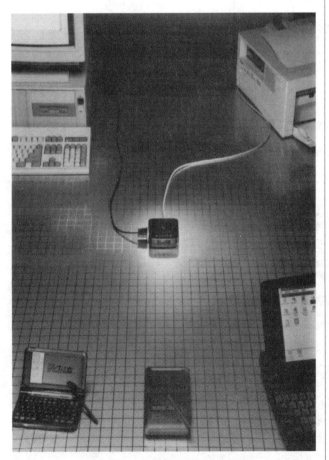

Figure 3-15. The ACTiSYS ACT-IR3S+ which can transmit to the PDAs surrounding it, and many more. (Courtesy of ACTiSYS)

chronization, backup, and network access using the appropriate transfer software.

The IrDA and multimode IR computer adapter attaches to the PC's serial port and gives you a wireless connection directly from IR based PCs and PDAs. The ACT-IR200L can transmit up to 9 feet while being powered just by the PC's serial port. It is also compatible with the IrDA standard and Sharp's ASK modulation scheme enabling communications with Sharp's Wizard and Zaurus, Apple's Newton, H-P's LX/Omnibook, IBM's Thinkpad 755™, and others. The ACT-IR200M is an IR wireless module interfacing directly with a notebook PC. The ACT-IR100M printer adapter allows IrDA PDAs and PCs to transmit to a printer while an ordinary PC uses the same printer. It attaches to the parallel port.

Troubleshooting Testers

The Genoa Technology SIR Tester allows you to quickly know whether an IrDA protocol PC or PDA can communicate with the standard IR serial interface. This instrument tests and indicates four critical IR activities:

1. If the IR peripheral passed or failed. If the peripheral test passes, a test page is printed.
2. If the computer or PDA passed or failed. This tester will emulate a peripheral by responding to a host computer's attempt to establish communications.
3. If it can send and receive IR frames. This helps test such devices as IR telephones.
4. If it can test for the presence of IR energy which eliminates the need for an expensive IR scope.

Puma Technology makes a similar device, although its data transfer rate is 1.15 megabits per second which would enable it to transfer a 20 page document in just one second. This device is the Fast IR Adapter and comes in the form of a PCMIA card for either a PC or a notebook and accompanying TranXit™ software for $299. There's a serial IR

Chapter 3

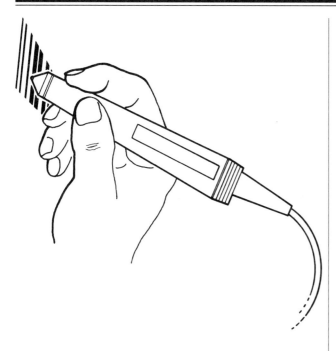

Figure 3-16. The physical appearance of an optical bar code wand as it reads a bar code.

adapter which plugs into the serial port of your computer and uses just PC port power, similar to the ACTiSYS systems.

Other IrDA Based Products

Attendants at car rental return lots wear a belt loop held small printer. This Seiko Instruments LTP-3000 Series 5-V print engine receives IrDA signals to wake up a controller prompted by a hand held bar scanning wand, our next subject.

Bar Code Technology

This accurate, easy to use, and inexpensive data entry and storage IR based scanning technology greatly improves productivity, traceability, and material management. It uses IR photodiodes and clever optics in their lend arrangement.

A History of the Bar Code

Two Drexel University engineering professors conceived this idea in the late 1940s. One professor, Bernard Silver, overhead his dean talking to a grocery store president lamenting how long the checkout counters took and asked for help.

The first idea read colored fluorescent dots with UV lights, but limited price ranges. Black and white bars posed uncertainties in the direction the light passed through the scanner. A bull's eye pattern of concentric circles in 1952 used lighting sources which were still too primitive. In 1959, ICs and lasers came of age. IBM jointly developed a bar code system and implemented it between 1970 to 1974. After technology finally caught up, delays stemmed from failure to agree on a standardized groceries code.

This technology extracts data from a medium by optical means. It usually is an optical scanning hand

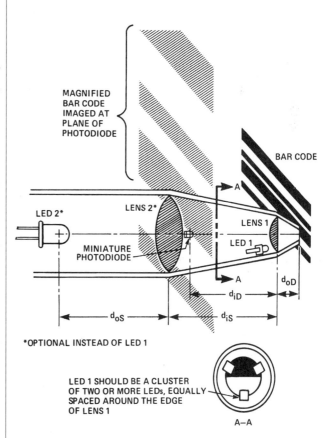

Figure 3-17. The interior lens and IR LED arrangement of a bar code scanning wand. (Courtesy of Hewlett-Packard)

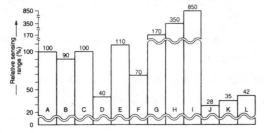

Difference in sensing range by sensing object (Applicable for diffuse reflective sensors)

A : Non-glossy white paper (standard)
B : Natural color cardboards
C : Plywoods
D : Non-glossy black paper (brightness : 3)
E : Glossy plywoods Natural color bakelite boards Acrylic boards (black) Vinyl leathers (red)
F : Vinyl leathers (gray)
G : Glossy green rubber boards
H : Aluminum boards
I : Reflectors
J : φ10mm steel rods with rust φ5mm brass pipes
K : Clothes (black)
L : Clothes (dark blue)

(*1) : In the above graph, the sensing range for non-glossy white paper is 100. This is the standard sensing range. The relative sensing range is a ratio against the standard sensing range. These figures are guidance only and will change according to the sensing object size and the type of photoelectric sensors.

Figure 3-18. A graph of the reflective qualities of some common objects judged against non-glossy white paper as 100%. (Courtesy of SUNX)

wand. (See *Figure 3-16*) A logic signal results from the difference in reflectivity of the printed bars and their underlying media (spaces). You retrieve this serially stored data by scanning in a smooth, continuous motion.

An Optical Scanning Wand

Figure 3-17 is an optical bar code scanning wand's interior IR LED and lens arrangement. This is a good example of reflective sensing with a coaxial lens arrangement. It might be possible to operate with no lenses; however, a lens greatly benefits the detected signal and its resolution in edge sensing applications like this.

The IR LEDs which irradiate the bar code should not be imaged (focused) at the code's plane. This would increase flux coupled to the IR LED; however, it could very well cause interaction between the details of the LED's image and the code. That is why LED 1 is a cluster. If you elect to use LED 2, instead of LED 1 (See *Figure 3-17*) notice it is imaged at the plane of LED 1. By focusing LED 2 at LED 1's plane, it defocuses at the bar code's plane and prevents interference.

The following equation defines the line resolution of the bar code as the diameter of the IR LED photodiode times the ratio of d_{iS} to d_{oS}:

$$1/d_{oD} + 1/d_{iD} = \text{line resolution}$$

d_{oD} is the distance from LENS 1's flat surface to the bar codes. diD is the distance from the wand's light emitter to LENS 1's flat surface. (See *Figure 3-17*) Notice the distance between the positions of LED 1 and LED 2. Minimize the ratio of d_{iS} to d_{oS}. That is, make it as small as the imaging requirements allow, which maximizes performance.

Judging System Performance

Two parameters describe a bar code system's performance. The first parameter is first read rate, the ratio of good scans to total scans attempted. Good

Figure 3-19. The Seiko MessageWatch™. (Courtesy of Seiko)

Figure 3-20. The Seiko Message Watch™'s two ASICs. (Courtesy of Seiko)

systems yield over an 80% first read rate. If not, one of three problems exist:

1. There's a poorly printed symbol.
2. The scanner's resolution is not well matched to the bar code symbols.
3. The system algorithm is very intolerant of errors.

The second parameter is substitution error rate, the ratio of invalid to valid characters entered into the database. Substitution error rate depends on:

1. The bar code symbology.
2. The printed symbol's quality.
3. The decoding algorithm. One extensively tested code is the 3 of 9.

A well-designed optical decoder using this system experiences far less than one error per million characters. Typically, a narrow bar is a logic 0 and a wider bar is a logic 1.

Special Purpose Bar Code Characters

Depending upon the bar coding symbols you select, the optical wand may require the presence of several special characters before it accepts a serial data stream. This adds symbology security; otherwise, you might scan reflective and non-reflective areas that just coincidentally have a valid bar code symbol, or sequence of symbols.

The start and stop margins, or quiet zones, are devoid of any characters or information. The start character is a special bar/space pattern identifying the start of a symbol. The stop character is a spe-

Figure 3-21. The Polar Accurex™. (Courtesy of Polar)

ters are also asymmetrical. This asymmetry allows interchangeable use of start and stop characters. This enables the optical decoder to differentiate between scanning in the forward and reverse directions. When you scan in the reverse direction, the decoder reorients the message characters to their correct order, prior to checksum calculation or message transmission.

Optical Properties of the Scanned Medium

The difference in the bars' and spaces' reflectivity determines the scanner's output. The illuminated area's small size enhances scanner sensitivity to printing flaws. This is far in excess of the naked eye and places much more stringent requirements on the printer's tolerances and inking regularity. The bar and space pattern determines contrast which affects reflectivity. You specify surface reflectivity at a specific optical wavelength and radiation pattern. Surface reflectivity is the amount of light reflected when an optical emitter irradiates the media surface. Ideally, it is between 70% and 90% of the incident light. White paper commonly achieves this.

SUNX, a domestic manufacturer of photoelectric sensing modules, uses the graph in *Figure 3-18* to quantify objects their reflective diffuse sensors sense. This graph uses non-glossy white paper as 100%. It references or judges other objects against this. Note in column "I" the 850% reflectors yield.

Figure 3-22. The Polar Vantage NV^{tm}.
(Courtesy of Polar)

cial pattern signifying the symbol ends. This character also signals to total a check sum, if your system uses one. Most bar code symbologies themselves define the check sum character. The bar/space pattern indicating the start or stop charac-

Figure 3-23. The Beep^{tm}. (Courtesy of Swatch Telecom)

Chapter 3

Figure 3-24. The Breitling Emergency™ watch and beacon. (Courtesy of Breitling)

The optical pattern of light leaving the media surface is the reflected radiation pattern. A shiny, or specular surface generates a narrow radiation pattern. A dull or matte surface yields a diffuse pattern. Avoid paper with a narrow radiation pattern. If this truly interests you, there are surface reflectivity meters made by the Macbeth Division of EG&G and by Photographic Sciences.

IR Technology at Its Best, & Its Competition

A New York City marketing research firm, Packaged Facts, estimates there is a potential $23 billion market for electronic wrist instruments. Merely referring to them as watches does them a grave disservice and insults their capabilities. This rapidly growing market already has a surprisingly vast breadth of electronic wrist instruments including pocket organizers, TV remote controls, pagers, and message transmitters/receivers. This wrist instrument market divides into 35% IR based products and 65% non-IR based products. Let's review representative examples of the non-IR based market first and then conclude by contrasting them to IR-based products. Non-IR RF products, due to their very natures of limited versus longer ranges, never overlap functions and therefore provide no real competition to IR based products.

COMPARING THE TECHNOLOGIES

Non-IR Based Wrist Instruments

The larger non-IR wrist instrument market almost exclusively uses RF in the form of low power se-

Figure 3-25. The Casio Triple Sensor™. (Courtesy of Casio)

71

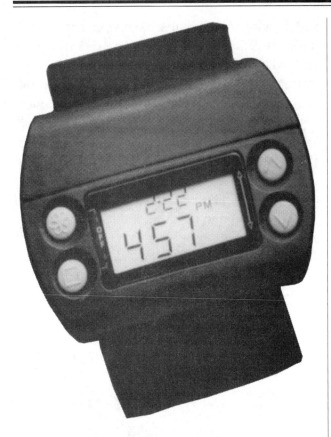

Figure 3-26. The GlucoWatch™. (Courtesy of Cyngus)

can designate any message urgent and decide when to respond if the message warrants the cellular charge. Voice mail can outdial your MessageWatch™ to let it know when you have a new message. If you are out of the service area, you can call your private message center for messages.

The MessageWatch™ retails from $80 to $130 with monthly service as low as $7.95 and an activation fee of $20. Voice mail fees cost from $5 to $14 per month. The actual time of the watch is always correct within a second since it updates itself 36 times daily to NIST's (formerly called the National Bureau of Standards) atomic clock.

lective FM frequencies. This is not disruptive to commercial stations primarily due to two practices:

1. Sharp transmission filter skirts ensure extremely little impact on the main channel in single path situations.
2. The error detection schemes are also flexible since, e.g., the MessageWatch™ corrects any short burst of errors associated with random noise or automotive ignition noise.

The Seiko MessageWatch™

In this category, the Seiko MessageWatch™ is by far the most sophisticated example and uses a subcarrier on the FM band. These also come in slightly different models. (See *Figure 3-19*) This wrist instrument is not only a watch, but also a message center which has its own phone number, and receives and displays 16-digit messages with the phone number's area code and extension. You

Figure 3-27. The Timex DataLink™ IR-based watch in a metal band. (Courtesy of Timex)

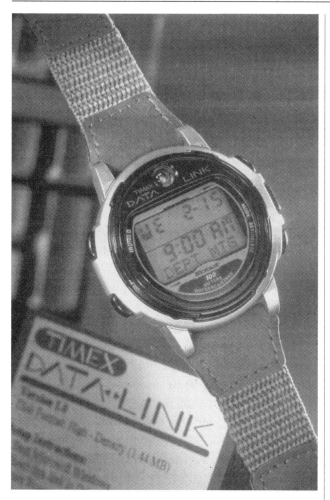

Figure 3-28. The Timex DataLink™ IR-based watch in a leather band. (Courtesy of Timex)

Message Selection

From the viewpoint of the user, this is its strongest feature. The variety and depth of messages are astounding. The MessageWatch™ displays the messages you receive on its 7-segment multi-digit LCD. A typical message might come from air pollution monitoring agencies. Its display indicates "PAr" (particulates), "CO" (carbon monoxide), "n02" (nitrogen dioxide), and "O3" (ozone) in the air. There are numbers to the right of these air indicators showing the pollutants' severity. Numbers below 50 are good, from 50 to 100 moderate, 100 to 200 unhealthy, 200 to 275 very unhealthy, and above 275 hazardous. The most common pollutant in Southern California is carbon dioxide while Northwesterners mostly observe particulate counts.

The Southwest Air Quality District reports for LA and Oregon's Department of Environmental Quality issues reports for the Portland area. The Puegot Sound Air Pollution Control Agency provides this data to Seattle and its surroundings.

Other service messages displayed include pro sports scores, weather forecasts, ski conditions, surf heights, state lottery numbers, and financial information. Due to the inherent limitations in a 7-segment display, some sports teams have unfilled segments which, without the table Seiko supplies, are indecipherable. Most team designators are comprehensible (e.g., the Anaheim Ducks in hockey use "dU.") The Seiko MessageWatch™ ski conditions are in four parts, one in each corner of the display. These are:

1. Freezing level (altitude).
2. Inches of new snow.
3. Base snow level.
4. Current temperature.

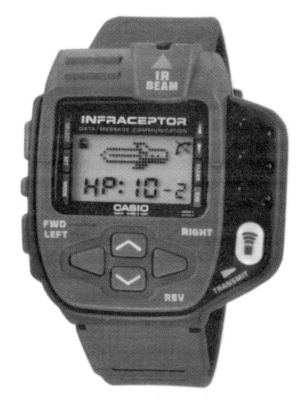

Figure 3-29. The IR-based Casio Infraceptor™. (Courtesy of Casio)

Financial enthusiasts receive NASDAQ (over the counter) closing averages, New York gold prices, Standard and Poor's 500 and Dow Jones averages (including volume). There is a special feature which activates during the day anytime the Dow moves 100 points. It immediately alerts you to the magnitude and direction of the change, as well as prime rate changes, an infrequent, but very significant occurrence. If requested, Seiko's customer service department also sets up MessageWatch™ prompts in Spanish.

Limited Coverage Area

The Seiko MessageWatch™ is an FM receiver of local FM station radio frequencies. It uses seven FM stations in the Seattle and Tacoma area with a coverage of 2.6 million people. It uses 19 FM stations in the LA, San Diego and Palm Springs area to cover 17 million people. Las Vegas has just recently come on board and the Netherlands is on board in Europe.

Reception

Since higher frequency FM reception is more limited than AM, when you leave the receiving area and/or enter some physical barrier (such as a tunnel), your MessageWatch™ loses contact with its radio signal. It then automatically searches FM frequencies for a new signal. After a prolonged extensive search, an "S" appears for Search ended. This frequency agile feature allows it to automatically tune to any local HSDS (Seiko communication's FM subcarrier High Speed Data System) equipped station on the FM band. This eliminates the need for FM frequencies dedicated just to traveler service stations. Two custom ASICs (application specific integrated circuits) accomplish all these amazing feats. (See *Figure 3-20*) Note these ICs reside on boards with identical shapes and sizes. The ICs' pads on top are insulating elements so the ICs don't short out to each other, or the watch's metal case.

The Polar Heart Rate Monitor

These wrist instruments are heart rate monitors made by Polar, a Finnish company, and developed by a Finnish cardiologist. Many exercise equipment stores carry them. The instrument consists of an elastic chest band which senses your heart beat and relays this data to the wrist instrument. There are seven different models of heart rate monitors ranging from simple instruments which just give heart rate, such as the Accurex™ (See *Figure 3-21*) to the Vantage NV (night vision) which provides a vast array of data and features. The Vantage NV™ (See *Figure 3-22*) transfers data through an RF link to a Windows driven PC for storage and further analysis.

There is a two-way RF communications link which permits electronic programming of one or more Vantage NV™ heart rate monitors which include user ID. You can transmit and receive raw heart beat data from any chest transmitter in real time form. The flexible advanced diary driven software package enables you to record history and plan future exercise sessions. The document generated gives multiple reports and graphs of heart beat and a scattergram playback of heart rate variability, which you can use to analyze your stress levels.

A Stylish Beeper

Swatch, the trendy art nouveau watch manufacturer makes The Beep™, an analog style watch and pager. Swatch claims this is the world's smallest radio receiver. (See *Figure 3-23*) The watch's band cleverly envelopes its tiny antenna and provides 5 mV sensitivity.

The original Beep™ was a tone-only pager capable of only transmitting four different sound signals. The Beep™ evolved into a small, easily readable display for incoming messages directed toward a youth oriented mass market. The Beep™ is activated by calling an 800 number. This is

MobileComm, a division of Bell South. There is an activation fee and a small monthly service fee. It is already under redesign to make its features appeal to the computer literate market. Its next generation will communicate via PCs or Macs through a modem and receive and store up to 10 messages of 80 alphanumeric characters each.

The Breitling Emergency™

This $4800 Swiss watch is for adventurers, such as mountain climbers or pilots, who could be in distress. This wrist instrument does not use FM but AM transmission, slightly above the FM band. The watch has two small antennas which extend from the watch and transmit at the aviation distress VHF frequency of 121.5 MHz. (See *Figure 3-24*) All aircraft operating under international civil-aviation rules carry a radio beacon set to 121.5 MHz, 243 MHz, or 406.025 MHz. Each watch kit comes with a test receiver and separate batteries. You can activate the Breitling Emergency™ for testing only. You do this with a test button on its underside which transmits a very weak AM signal at 121.5 MHz for 0.75 seconds duration every 2.25 seconds to the test receiver.

In an actual emergency situation, you activate the Breitling Emergency™ when you extend the antenna, the one on the lower right in *Figure 3-24*. This snaps the cap. It breaks away and the 17 inch main antenna emerges as an accordion-like thin copper strip. You can also extend the other antenna on the left; however, at sea, this is futile since the water reflects this energy. This unique signal indistinguishably identifies itself by transmitting a Morse code "B" every 60 seconds.

After rescue, you can cease distress transmissions by either cutting this antenna where it enters the watch or wrap it around the watch to short circuit it. The watch is a one time use design. After deploying the main antenna and the transmitter you can return it to the factory, with a documented legitimate justification of its use, and they will restore it to its original condition free of charge. In case you crash, or sustain a jarring fall which smashes the watch portion of this instrument, it may stop; however, don't assume the transmitter won't work. It is a separate function of the watch.

Under ideal conditions (transmitting from a mountain top), an aircraft 250 miles away flying at 33,000 feet altitude can detect this signal. The peak transmission power is 30 mW. The following equation describes the Breitling Emergency™'s actual range:

$$R = 0.8 \text{ to } 1.1 \ (h_t^{1/2} + h_r^{1/2})$$

R is the reception or transmission range in nautical miles. The *ht* is the height of the transmitter in feet and hr is the height of the receiver in feet. VHF frequencies tend to have the limiting factor of a line-of-sight range. Another limiting factor to transmission is static interference which increases at night, at lower altitudes, on land and during summer.

Two Specialized Wrist Instruments

The next brief descriptions describe two non-transmitting and non-IR, but very useful wrist instruments. The Casio Triple Sensor™ (See *Figure 3-25*) costs $250 and contains three distinct instruments:

1. A compass.
2. An altimeter and barometer.
3. A thermometer.

Activating the altimeter mode enables measurements for the first five seconds. Subsequently, it takes measurements every two minutes at 20 feet increments with a range of 0 to 13,120 feet (not accommodating Pike's Peak which is 14,110 feet above sea level). The barometer function:

1. Automatically takes air pressure readings every two hours.
2. Displays the current reading.

3. Stores and graphs the last 26 hours of barometric readings to predict weather.

The last non-IR wrist instrument is the GlucoWatch™ from Cygnus. (See *Figure 3-26*) This company pioneered nicotine and estrogen patches. The GlucoWatch™ uses a GlucoPad™. Patients with severe diabetes, a disease caused by the pancreas not producing enough insulin, have to monitor their blood sugar levels as often as every half hour through an invasive, often painful, pin prick to obtain blood for testing. Now, patients can wear the GlucoWatch™, a wrist watch that continuously monitors and calculates trend analyses in blood sugar level. Its alarm sounds and awakens for both hypo- and hyperglycemia. The GlucoWatch™ non-invasively extracts glucose molecules through intact skin. It uses low current levels and a proprietary electro-osmosis process.

The disposable GlucoPad™, which adheres to the skin, and connects to a wrist sensor under this pad, collects these glucose molecules daily. The collected glucose triggers an electrochemical reaction with a reagent in the GlucoPad™ which generates an electric current. A reagent is often an ultra-pure chemical which promotes a process. The sensor measures the electrons and a custom application specific integrated circuit (ASIC) in the GlucoWatch™ equates these currents to the level of glucose in your blood. The GlucoWatch™ costs $400 and lasts approximately three years.

Optical-Based Wrist Instruments

No doubt, you can remember scorning the first calculator wrist watches with their small keypads. These forerunners of today's sophisticated wrist instruments avoided using IR, simply for the lack of a source. PCs were not that prevalent then. The PCs which did exist didn't use IR, quite unlike today.

The Timex DataLink™

This watch comes in the model 70 series and the more recently introduced model 150 series. The DataLink™ receives data from your PC monitor's CRT through an optical wireless path. Starting with the 70 series, these come in five models with different bands and case housings, all just different cosmetics housing the same electronics. (See *Figures 3-27* and *3-28*) These show both metallic and leather bands respectively. You can't send data from the watch to a PC since it is only an optical receiver of data. It does not use optical PC ports and typically runs three years on its battery. The watch's optical sensor reads flashing light bars from your PC monitor. It does not work with LCD, STN, active matrix laptop displays, or Apple PCs.

The Timex DataLink™ watch accepts serial data from your PC's VGA or CRT and places it in your watch to remind you of an appointment, to-do lists, telephone numbers, and even data you obtain over a LAN. This could link and synchronize a group with a common purpose, such as employees within a company's department. The watch's accompanying software comes on a more modern 3.5" disk, but if you only have a 5.25" 720 kilobyte disc drive, just call 1-800-FOR-TIMEX and they'll send you that type disk. It requires a PC that is a 386 or newer and Windows 3.1 or later. You'll also need 4 MB of RAM, 2.5 MB of free hard disk space, and a mouse. If you have Microsoft Office's Schedule+™, this will greatly facilitate data transfer.

The watch has five buttons around its face: NEXT, MODE, SET/DELETE, PREVIOUS, and INDIGLO. The NEXT button displays the next entry in most modes, and allows you to depress this button to peek at your next appointment. The Alarm mode displays the next alarm. The MODE button displays the next mode. The SET/DELETE button enters the Time or Alarm Set modes and marks or deletes entries into other modes. The PREVIOUS button displays previous entries in most modes. The INDIGLO button illuminates the watch so you can read it in the dark.

Sending Sample Data

To familiarize yourself with the database downloading function, you can send a prepared set of sample data from your computer to your watch. Your software manual has instructions for you to install and start the Timex DataLink™ communications program. One of the first options you see under the heading "Menu Bar Functions" is **Send Sample Data.**

Communication Mode

In the Communication mode, the Timex DataLink™ watch receives — directly from your monitor screen — information you select from databases in your PC!

Setting up the Transmission at the PC

The DataLink™'s PC software booklet enclosed with the watch provides information on how to operate the downloading software, and the database documentation for constructing the files in your computer. Once installed, you can send data to the DataLink™.

Sending the Data to the Watch

Press MODE until you reach the Timex DataLink™ Communication mode. The watch display reads COMM READY. Hold the watch (while on your wrist is fine) in front of the monitor from 6" to 12" away and as steady as possible. The system has a large tolerance for variations, but these are ideal conditions. Start the transmission by pressing ENTER or clicking the mouse. The transmission of the data is indicated by a pattern of white lines on a black monitor screen. The monitor will initially display a static pattern of lines for 5 seconds to allow you to position the watch. The watch will beep steadily when it is properly aligned. The monitor lines will then flash, indicating that data is being sent; the watch will sound periodic beeps to indicate that it is receiving the data. The transmission is completed when the PC screen returns to the previous display. The watch display will then show one of the following five messages:

COMM DONE, DATA OK: All data successfully received and stored. A single beep tone also sounds to acknowledge the successful transfer of data.

COMM ABORTED: You terminated the download by pressing any button on the watch.

COMM CEASED-RESEND: May mean you moved the watch so it could not clearly "see" the monitor screen. This or any other error message is accompanied by a varying tone similar to that sounded by a European emergency vehicle.

COMM ERROR-RESEND: Data was received but not understood by the watch. Repeat the transmission process.

COMM ERR #, SEE HELP: Make a note of the error number. Repeat the download procedure. If the error recurs, call Timex Technical Support and give them the error number. If communication fails, all database modes are unavailable in the watch (though still available in the database in your PC). If you choose to download no data for a particular database, that database mode will be unavailable in the watch.

Calibration

If you have tried to download sample data several times in succession, the software will recognize that a problem may exist, and will start the program to calibrate the transmission software. Follow the instructions on the screen, to adjust the pattern of lines until your watch responds with a series of beeps indicating that the synchronization is successful. Then execute your download. You can also select the calibration option whenever you wish; this is explained in the software manual.

DataLink™ Software

After loading the software and performing the communications test you can start sending data to the watch and doing other exciting tasks. At the bottom of your PC's screen you'll see a button labeled SEND TO WATCH. Clicking on this button sends the currently selected data to your watch. You can simply select this button at other times to send the watch data you may have designed yourself. A gas gauge type icon in the center of the screen tells how much of the possible data of the watch's memory capacity (in %) you have sent.

Adding Entries to the Database

If, for example, you have selected Get Appts from MS Schedule+ from the drop down menu on the Main Menu screen, this database area will not be available. To enter appointments, simply place the mouse pointer in the appropriate box and type in the relevant information. From the Edit menu item, you can select:

1. Cut.
2. Paste.
3. Copy.
4. Insert.

These selections work exactly as MS Windows works.

Importing Data From Other Databases

You start by clicking on **File**, then **Import**. In the dialogue box which appears, select the Timex DataLink™ database which will receive the data, Appointments, Phone Numbers, or To-Do Lists, and specify the filename of the database from which you took the data. By selecting either Append (the new data is added to whatever is already in the database) or Replace (the new data replaces what is already there) you can control how data is received.

The source files must be in CSV, or comma-delimited format. This means fields in each entry are separated by commas, name, number, message, etc. Most databases are stored in this format. Those which are not stored in this format usually export the data into a new file in the CSV format.

You may want to select entries to be imported since each Timex DataLink™ file holds a maximum of 100 entries (200 for phone numbers). **Import** works with larger databases, but will simply import the first 100 (or 200 phone numbers) entries it encounters. If you have selected Append, this allows you to import as many records as memory capacity allows.

The 150 Series of DataLink™ Watches

The new model 150 DataLink™ watches place twice as many data words (two 10-stripe patterns, instead of just one) on the CRT. This is why their data transfer rate is twice that of the model 70. Now, let's explore the difference in these watches by discussing everything the 150 does. The model 150 series (See *Figure 3-29*) comes in four styles and has all the features of the model 70 series DataLink™ watches, plus the following:

1. It transmits at twice the speed of the model 70.
2. It has a stopwatch and either a preset countdown timer or an adjustable countdown timer.
3. The Microsoft Windows Schedule+™ is built in and comes with optional WristApps™ software which allows you to send additional information to your DataLink™ model 150 series watch.
4. It has twice the memory, which allows 30 to 40 word notes and storage of up to 150 messages.
5. It displays a 31 character scrolling message line.
6. The "Make A List" feature is expanded to allow longer and prioritized lists.

7. There are a variety of high or low pitch tones you can select. This makes the beeps easier to hear and distinguish from other watches. Alarms and time settings do not affect memory capacity.
8. It instantly displays preprogrammed phone numbers from a touch.
9. It stores special events, such as anniversaries.
10. It automatically adjusts its time for Daylight Savings Time and has two time zone settings, and those of over 100 major world cities. It also displays either 24 hour military or normal 12 hour time.

THE MODEL 150 DATALINK™'S THEORY OF OPERATIONS
Selecting the Light Detector

Both series of watches have virtually the same basis of operation. The model 150 has added features doubling both its memory size and data transfer rate. Appropriate light sensors for such a design would be either innately fast switching photodiodes, which require external amplification, or a moderately fast switching phototransistor which provides its own amplification. However, the very high gain Darlington pair phototransistor is inappropriate due to its inherently slow switching time. Therefore it can't accommodate the PC monitor's fast light pulses. A rather ordinary silicon phototransistor, the Sharp PT370, ended up as the DataLink™'s photodetector. Its spectral response skews toward the visible spectrum's lower end (red). It is an optimized comprise of three considerations:

1. Its tolerance of a DataLink™ owner's arm movements.
2. Its ability to operate at a distance of about a 6" to 12" from the CRT.
3. Its ability to reject ambient light, especially fluorescent light with its multiple irregular spectral peaks.

Later in this book, we will examine an evaluation board using the Texas Instruments TSL230 opto-

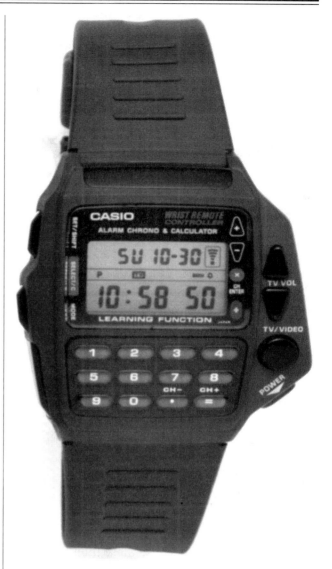

Figure 3-30. The IR-based Casio Wrist Remote Controller™. (Courtesy of Casio)

electronic IC. One of the experiments deals with light synchronization. It samples at exact intervals and nullifies the AC line frequency's 60 and 120 Hz components' adverse effects. This ensures optical sensed data is virtually impervious to this light source's adverse effects. The PML's (Physical Media Layer) circuitry processes and conditions its input signals (short light bursts) by high pass filtering so it ignores the fluorescent lights' effects. The AGC (automatic gain control) circuit (See *Figure 3-30*) minimizes the effects of your watch's distance from the CRT. Its varying signal power

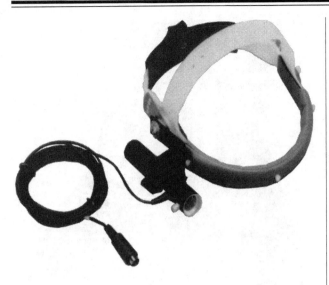

Figure 3-31. An IR-based head activated PC entry device and controller. (Courtesy of Design Technology)

is inversely proportional to the reciprocal of this distance squared. This circuitry also converts the RZ signals into an NRZ format.

Data Characteristics and Principle of Transmission

Data is one pixel wide and one scan width long; its period is uniform because these intervals don't change. The real difficulty is trying to accommodate every CRT based PC monitor on the market today. These monitors have a vast span of vertical scan rates, phosphor types, and different reaction times. As an example, Panasonic's ultra-high resolution PC monitors (1,440 x 1,800) use a P21 phosphor type and their vertical scan rates vary from 50 to 160 Hz.

A good analogy of how the DataLink™'s transmissions work is ship-to-ship light messages. They use a narrow radiation pattern strong light source with a mirrored surface reflector in the housing. The signalman's Morse Code-like light transmissions come from flipping louvered shutter type flaps which cover the light source and either pass, or quickly block transmissions.

After manually entering keyboard data into the PC, hold your DataLink™ watch's face from 12" to 18" away from the CRT and click the START button. The PC downloads this data to your DataLink™ through short pulses with active-low bits of alternating white or black images on the CRT. White corresponds to a logic 0 and black represents a logic 1. This is not full duplex two-way synchronous, but rather unidirectional asynchronous optical data transmission. Ten horizontal stripes fill your PC monitor's CRT. These framed 8-bit bytes are the middle eight stripes. The highest and lowest stripes are the start and stop bits respectively. The DataLink™'s integral phototransistor reads this stripe data and notifies you when the download ends.

The PML

The Physical Media Layer (PML) describes optical transmissions of the Timex Communication Protocol (TCP). This description requires a basic understanding of CRT operation. The following description uses just a black and white CRT to simplify matters. To illuminate one horizontal row, the raster scan sweeps from the CRT's left to right edge. (See *Figure 3-31*) To illuminate the entire CRT, the raster scan makes one left to right horizontal sweep, followed by a retrace, and starts again. This time though, it sweeps one row lower until it reaches the lower right corner. (See *Figure 3-30*) After this upper left to lower right corner sweep, it recycles. If there is a logic 0 (black image), the CRT lowers its electron beam acceleration potential to purposely not illuminate that row. *Figure 3-32* shows a repetitious alternating binary 1 to binary 0 pattern.

The TCP

The TCP transmits asynchronous serial RZ (Return to Zero) pulses in the 10-stripe pattern with the Least Significant Bit (LSB) first. (See *Figure 3-33*) Generally, the time (period) for one scan to travel from the beginning of one bar to the end of the next bar is approximately the reciprocal of the

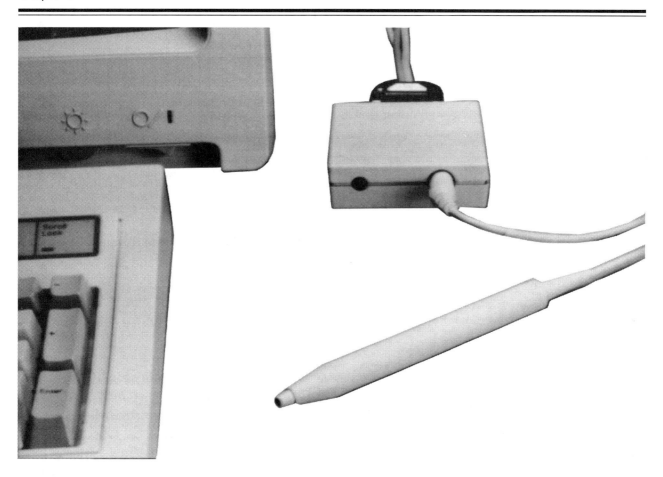

Figure 3-32. An IR-based PC light port and scanning pen. (Courtesy of Design Technology)

baud rate. Since the period equals several scan lines it differs on monitors running varying video systems with different scan rates. The principal parameter though is the time it takes between the start of successive bars. It is not the physical distance between bars since the DataLink™ does not image the screen. This is why selecting the actual positions of the bars is so important for avoiding any possible conflicts between the horizontal and vertical retrace intervals. Each refresh of the screen must contain a new set of data words, or be left blank. A lack of new data on the screen with each refresh forces the software to re-read the screen as many times as needed until this condition changes.

Pulse stretching inverts and converts the "pulsed" RZ signal into an NRZ format which compensates for any timing error instinctively characteristic of using an integral number of scan bars between each bar position. (See *Figure 3-34*) This final portion of circuitry (See *Figure 3-30*) makes the signal acceptable to a standard UART (universal asynchronous receiver-transmitter).

Limitations of Data Rates

The number of horizontal lines which can fit on the CRT screen, multiplied by the vertical scan refresh rate, determine the data rate. This limits the number of data words from one to three. Typically though, no more than two data words will ever appear on the CRT. The typical vertical refresh rate for PCs varies from 60 to 80 Hz; however, these rates can exceed these limits, as our previous example showed. This factor limits the data rate from one to three times the vertical scan rate. Using both 60 to 80 Hz for our example, this translates into 60 (60 x 1) to 180 (60 x 3) data

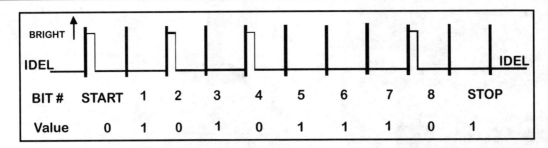

Figure 3-33. The TCP transmits asynchronous serial RZ pulses in the 10-stripe pattern with the LSB first.

words transmitted per second. It may logically seem to you the maximum data transfer rate should be 240 (80 x 3). However, this in unattainable because faster CRT refresh rates do not allow enough time to display three whole words.

The Casio Infraceptor™

This IR based wrist instrument retails for $69.95 and is more than just a data bank for storing phone numbers. Digital data travels bi-directionally via an IR beam, note the IR BEAM and up arrow designators in *Figure 3-29* indicate the IR source. You can't block, reflect, deflect, or impede this is line-of-sight optical energy IR beam and obtain proper operation. Also, you can't download data through a PC to the watch as the Timex DataLink™ does. But, the Infraceptor™ goes beyond the traditional pocket organizer which stores to-do lists and appointments. This wrist instrument stores 10 preprogrammed messages which you can both send and receive from an identical watch or other Casio IR wrist instrument.

There are two games you can play against yourself or against another wearer of an Infraceptor™ wrist instrument. The object of the Tower Master Adventurer™ game is to overcome various obstacles and find your way through trick gates, making your way to the tower's top. Once there, you fight to take possession of the Tower Master Orb. If successful, you save the world from demons. You can select your weapon from among:

1. A bow and arrow.
2. A whip.
3. A sword.

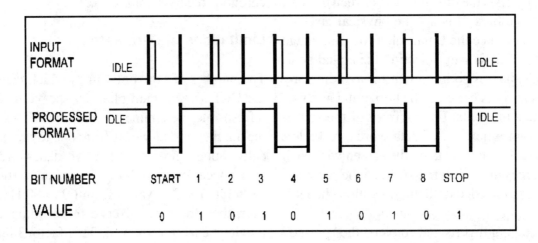

Figure 3-34. Pulse stretching inverts and converts the "pulsed" RZ signal into an NRZ format which compensates for any timing error instinctively characteristic of using an integral number of scan bars between each bar position.

Closely examine *Figure 3-29* to observe the bow and arrow in the upper right hand corner. If you lose, the demons throw you into an underground dungeon.

This wrist instrument has a built-in stopwatch, a daily alarm, and a telephone directory. This allows you to store names, numbers and memos for up to 10 people. The memory holds a maximum 43 characters and 380 data bytes. The Infraceptor™ is accurate to within 15 seconds a month and typically runs 15 months on its lithium battery.

The Infraceptor™, like other wrist instruments in the Casio line, interacts with its diaries. One diary is the JD-8000, the so called Hoop Commander™. It allows you to play basketball against your friends, the computer, or David Robinson. There is a player profile listing names, teams, positions and statistics of your favorite NBA players. There is also a scrolling world map, full function calculator, and a diary with secret password. This fold up hand instrument beams an IR message and interacts with other Casio wrist instruments.

The Casio Wrist Remote Controller

You won't easily misplace this $130 IR based TV remote control since it resides on your wrist. (See *Figure 3-30*) It allows you to run VCRs and cable boxes through its IR beam. You can select any one of the following five cable boxes:

1. Jerrold.
2. Panasonic.
3. Philips.
4. Pioneer.
5. Scientific Atlanta.

The Wrist Remote Controller turns these boxes on or off via its IR remote commands. You can program this wrist instrument for a broad variety of consumer electronic products from a large number of different manufacturers. If you can sacrifice the calculator function and some of this wrist instrument's high end programmability, you can buy the Casio model CMD-30 for about $80.

IR Based Products Aiding the Physically Challenged

The head activated IR transmitter in *Figure 3-31* enables quadriplegics to enter data and control a PC. This product's manufacturer, Design Technology, also makes a PC light pen port for IR scanning purposes. (See *Figure 3-32*)

Chapter 3

Chapter 3 Quiz

1. The IrDA standard's driving force was:
   A. The desire to increase transfer speeds of IR peripherals.
   B. The desire for greater reliability in IR based devices.
   C. The desire to have standardization in IR peripherals.
   D. None of the above.
2. In 1996 the IrDA standard reached:
   A. 4 Mbits/sec.
   B. 115 kbits/sec.
   C. 9.6 kbits/sec.
   D. None of the above.
3. The IrDA:
   A. None of the below.
   B. Vigorously competes with RF (Radio Frequency) technology.
   C. Stresses communications above connectivity.
   D. Requires FCC approval.
4. The newer IrDA:
   A. Has a shorter wavelength than the old system.
   B. Has a longer wavelength than the old system.
   C. Can still converse with most older IrDA systems.
   D. Both A and C.
5. The new IrDA system:
   A. Makes trade-offs in reliability vs. cost.
   B. Makes trade-offs in range vs. power.
   C. Uses more power.
   D. All the above.
6. What factor primarily determines the quantity of light reaching the IR detector LED?
   A. The shape of the IR LED's housing
   B. The angle of light's arrival
   C. The wavelength of the light
   D. None of the above
7. Which of the following have slower modulation rates?
   A. The IBM Thinkpad 755^{tm}
   B. The H-P 95LX
   C. The Sharp and Newton PDAs
   D. All the above
8. A series limiting resistor limits what in an IR LED circuit?
   A. Noise
   B. Instantaneous peak currents
   C. Reliability
   D. None of the above

9. You can make your own IR filter by:
   A. Purposely exposing ASA/ISO 100 color film.
   B. Developing it without making prints.
   C. Double layering the exposed and processed negatives.
   D. All the above steps, in order.
10. What must a manufacturer do before displaying the IrDA logo?
    A. Thoroughly test the software
    B. Thoroughly test the hardware
    C. Thoroughly test the software and hardware
    D. None of the above
11. The Bit Error Rate (BER) software allows reliability testing of the IrDA physical layer at various:
    A. Angles of transmission.
    B. Distances.
    C. Power levels and adverse lighting conditions.
    D. All the above.
12. There is hardware allowing you to with non-IrDA compliant peripherals, T or F?
13. There is troubleshooting hardware allowing you to send and receive IrDA frames to help analyze IR telephones, T or F?
14. Who first conceived the bar coding idea?
    A. Two professors at Drexel University
    B. Two professors at Purdue University
    C. Two professors at Harvard University
    D. None of the above
15. An optical scanning IR wand's lens causes the emitting LED to directly focus on the plane of the bar code, T or F?
16. The two new ICs which compete in the car security market are made by:
    A. RCA.
    B. Telefunken.
    C. Motorola.
    D. Intel.
17. These two car security ICs do not invade the IrDA's main realm, which is:
    A. Long range connectivity.
    B. Short range communications.
    C. Short range line of sight connectivity.
    D. None of the above.

Chapter 4
Optocouplers

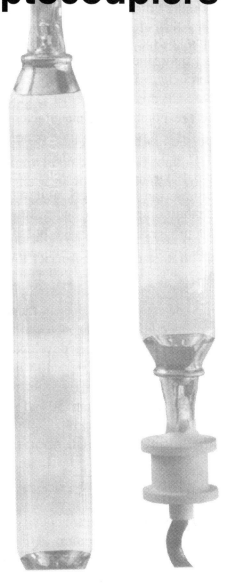

Chapter 4
Optocouplers

As their interchangeable names imply, an optocoupler or optoisolator couples signals but isolates grounds from one circuit to another. The ability to have different isolated grounds ensures hospital patients being monitored are safe from potentially hazardous shocks while virtually eliminating ground loop currents. The action of this device is very similar to that of an electromechanical relay which has a coil, which once energized, has its contacts snap shut. This completes the circuit and allows current to flow.

Optocoupler History

Optocouplers have advanced from crude laboratory curiosities in the late 1950s to the first commercially available units, which were light sensitive SCR optocouplers in 1964 made by GE. The next year optocoupler manufacturers first used the liquid phase epitaxial process. In 1971, optocouplers first used glass dielectrics. In 1975, the first GaAs IR LED coupler demonstrated a four million hour half life operating at 55 °C. Then the reliability increased to six million hours in 1981. The progress since then has been in increasing reliability, shrinking sizes, reducing manufacturing costs, and nullifying CTR degradation.

Photodetectors don't present manufacturers with as many problems as do photoemitters. That is why, as previously stated, their development lagged photodetectors by about 20 years. Silicon photodetectors have a wide response from about 700 to 950 *nm* with peaks between 750 and 900 *nm*.

An Optocoupler's Components

An optoisolator or optocoupler has three elements:

1. An IR photon emitting device.
2. A transparent medium.
3. An IR photodetector "tuned" to the same frequency as the emitter.

The emitter/detector combination is as closely matched as possible since admittedly, e.g. Motorola uses an inexpensive-to-manufacture 940 *nm* emitter which is not precisely matched to the tolerant photodetector. The emitter may be an incandescent, neon lamp or an LED. We'll discuss the modern LED integrated (IC) optocoupler. The medium may be a transparent insulation, optical fiber, air or glass. The photodetector may be a photodiode, phototransistor, photo-FET or some integrated combination of these. This obvious latitude in emitter medium and detector combinations produces an extensive array of different input, output, and coupling characteristics.

An Optocoupler's Emitter

An optocoupler's main concern for the emitter is coupling efficiency with the photodetector. You desire a low resistance, so a GaAs based optocoupler is the best choice. A low forward voltage is also important; however, this is not as crucial as the gain and bandwidth. Optical considerations of an optocoupler are drastically different from an LED. LEDs have annular (ring) shaped emitting patterns around a centered bonding pad. This gives a large ratio of apparent-to-actual emitting area. An optocoupler has as small an emitting area as possible. There is also an offset bonding pad on the IC substrate which ensures against shadowing of the emitting area and results in close photodetector coupling.

An Optocoupler's Photodetector

You can use photodiodes as photodetectors but this requires adding an amplifier for adequate drive. It is costly and inefficient to add discrete external amplification. Therefore, integrated (IC) optocouplers have photodiodes built on board. There are two methods to accomplish this:

1. Use a phototransistor with its collector-base junction as the light detection region.
2. Use a photodiode with a separate transistor on the IC to amplify this photodiode's photocurrent.

Despite being less expensive to fabricate, the phototransistor has inherently slow switching speed and linearity problems. (See *Figure 4-1*)

The principal contributor to non-linearity in optocouplers is the collector current flowing in the collector-base junction. This reduces the collector-base depletion region which likewise decreases responsivity. Collector current does not flow in the photodiode/transistor combination, even if the photodiode's cathode connects to the collector. This is referred to as an optocoupler's "phototransistor connection." (See *Figure 4-1* again.) You obtain optimal linearity by maintaining a fixed voltage across the photodiode.

An Optocoupler's Response Time

This is inherently slow due to excessive junction capacitance. However, this junction, by necessity, must be large to capture adequate photons. The Miller effect amplifies this junction capacitance of approximately 20 pF. The Miller effect, though, applies to an extremely small capacitance, 0.5 pF when you add a photodiode, see the illustration to the right in *Figure 4-2*. Making this junction thinner does reduce capacitance with no adverse effects. You obviously have no control over these

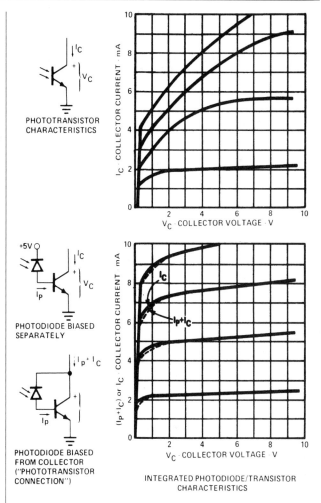

Figure 4-1. Linearity of photodiodes vs. phototransistors. (Courtesy of Hewlett-Packard)

internal optocoupler connections; however, they don't comprise as great a portion of capacitance as do the pins on this IC's case and the external connections. New technology's use of surface mounted components, with their diminished sizes, further reduce this major source of overall capacitance.

Using Optocouplers

Virtually all modern optocouplers are integrated, not hybrid devices. A hybrid device has microcomponents residing on a monolithic substrate which serves as a PC board, or interconnecting medium, for these components. Therefore, the emitter voltage must always be at or below the output

Real World Optocouplers

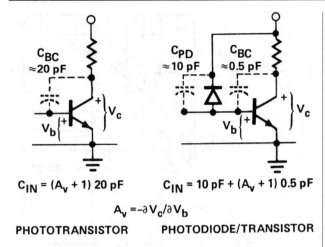

Figure 4-2. The input capacitance of photodiodes vs. phototransistors. (Courtesy of Hewlett-Packard)

There are many types in common use. One is the single transistor with photodiode anode connected to the base and the cathode separated for a reverse bias connection. Another is a phototransistor pair called a Darlington pair. (See *Figure 4-4*) The split Darlington configuration, characterized by bringing out the upper transistor's collector, allows the output (collector) to drop to a lower V_{CE} during saturation. This lower V_{CE} provides better noise immunity which is important in digital switching applications.

The base of the second transistor (pin 7), in *Figure 4-4*, is available for power strobing and resistive bypassing for increasing switching speed. This split Darlington optocoupler arrangement, with its high gain, makes it useful in many low power applications. An example is a gain element in a closed loop where you must isolate the input to the main amplifier from the reference comparator. Another example is in an analog power supply regulator. But since the first transistor in this configuration

at any other point on the optocoupler. *Figure 4-3* shows correct and incorrect connections. If the emitter drives some above ground point, such as another transistor, and if you use bypass resistors to enhance speed, you'll need to exercise care not to ground these resistors to the optocoupler output transistor's base. You also can't float this transistor's emitter.

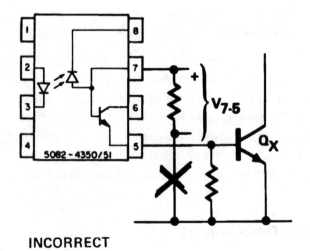

INCORRECT

RESIDUAL CHARGE ON BASE OF Q_X MAY MAKE $V_{7-5} < 0$ AND CAUSE EXTREMELY SLOW OPERATION

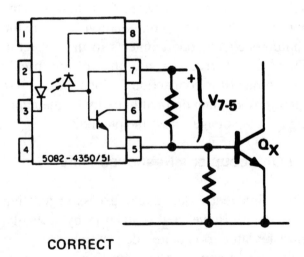

CORRECT

V_{7-5} CANNOT BECOME NEGATIVE

Figure 4-3. Connecting to the base bypass resistor with the emitter above ground. (Courtesy of Hewlett-Packard)

Chapter 4

is inaccessible, it is not very useful in analog applications.

Optocoupler Data Sheets

When you decide to select an optocoupler you'll first examine the data sheet and see five salient parameters, in descending importance, as follows:

1. Isolation (Common Mode Rejection, or CMR).
2. Insulation (maximum V_{I-O}).
3. Response speed (modulation bandwidth, propagation delay).
4. Reverse coupling (ground looping).
5. Forward coupling (Current Transfer Ratio, or CTR) or fan-out driving capability.

Isolation

The primary reason to use an optocoupler, or optoisolator, is in isolating or separating the input from the output. You may do this by optical, magnetic, or electrical means. The best means of coupling, though, is optical because the photons that carry the differential mode signal do not carry any charge nor do they need a magnetic flux to carry out their movement. There are only two methods by which this common-mode signal reaches the output:

1. Modulating or varying the input current.
2. By stray capacitance.

You can eliminate the first way of accomplishing flux movement by proper impedance balancing. (See *Figure 4-5*) Isolation characterizing is therefore a function of stray capacitance coupling.

Insulation

While this may imply, in some electrical devices, operation up to the point of breakdown, this is not true with optocouplers. The "corona", which occurs in optocouplers, is a partial discharge taking place within the optocoupler's insulation materials. Irregular distribution of electrical fields within

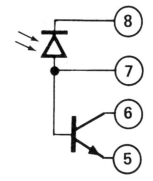

SINGLE TRANSISTOR AMPLIFIER

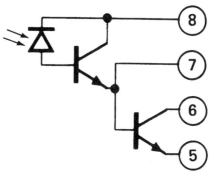

SPLIT-DARLINGTON AMPLIFIER

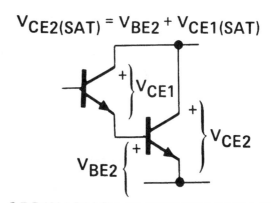

ORDINARY DARLINGTON AMPLIFIER

Figure 4-4. Analog type optocoupler photocurrent amplifiers. (Courtesy of Hewlett-Packard)

the insulation generate localized fields across what are called microvoids. They reach potentials at which breakdown occurs with a resulting partial discharge. You may see this expressed as CIV, or Corona Inception Voltage, on a data sheet.

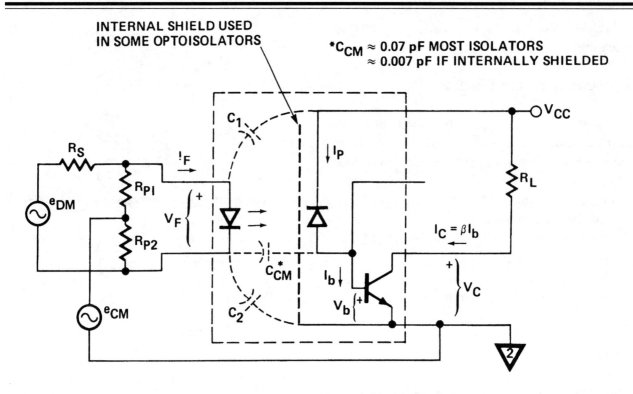

Figure 4-5. An optocoupler's input resistances and capacitances. *(Courtesy of Hewlett-Packard)*

Response Speed

The speed of operation is highly dependent upon both the mode of operation, analog or digital, and on the optoisolator you select. Peaking and feedback improve speed. Analog operation, just as with a transistor, requires the optocoupler to operate in the active region. This means the output (collector) neither saturates nor shuts off. In the circuit of *Figure 4-5*, the selection of input resistors R_S, R_{P1}, and R_{P2} must allow I_F to make $V_C \ll 0 \ll V_{CC}$. In the digital operation of an optocoupler, the low state, I_F forward current in *Figure 4-5*, must be large enough to bring the output voltage V_C well below some defined threshold. The high state current, I_F, must also be low enough to allow V_C to rise well above that threshold. Speed is the measure of how long it takes for the output to react to a change on the input. You characterize speed by propagation delay and the data rate.

Propagation Delay

Just as is true with a logic gate, optocouplers use this parameter in the same manner. They define the time it takes for a logic state to change the output after the device's input changes. There are more parameters which influence propagation delay in an optocoupler; however, the specific application most largely influences this. If a circuit draws a ments. This is a JEDEC (Joint Electronic Devices)

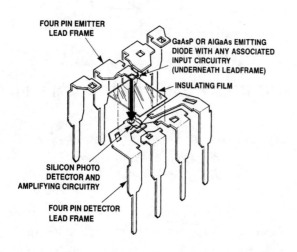

Figure 4-6. A DIP packaged optocoupler. *(Courtesy of Hewlett-Packard)*

Chapter 4

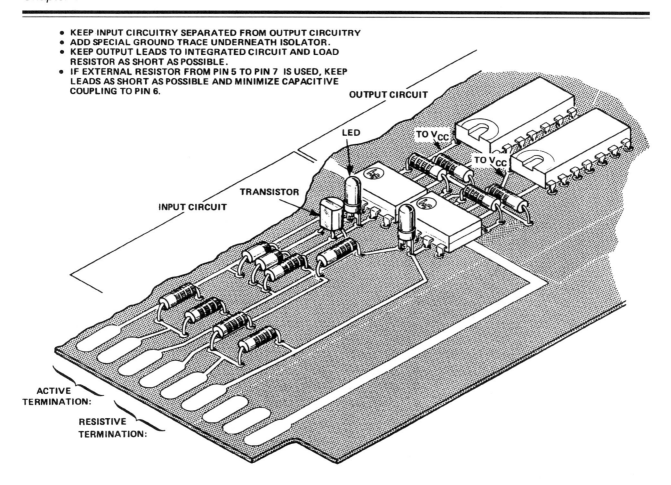

Figure 4-7. Proper PC layout for optocouplers. (Courtesy of Hewlett-Packard)

large collector current (output), it more deeply saturates the transistor and increases the transistor's storage time. Data rate is a measure of the maximum frequency of transmission of a square wave before it distorts.

Reverse Coupling

One of the main reasons to use optocouplers is to prevent flow of ground loop current. The very high (10^{12} Ω) impedance between the input and output in a typical optocoupler virtually guarantees elimination of DC ground loops. There are some minute AC ground loop currents due to a typical 0.07 pF C_{CM} (See *Figure 4-5*); however, in applications in which this reaches 1 pF it is usually unacceptable. To remedy this you will have to select another type optocoupler with a greater separation between the input and output. This usually entails using a more expensive optocoupler with either an actual lens or optical filter in between input and output optoelectronic elements.

CTR Trade-Offs

CTR is the ratio of input to output current and, the faster the optocoupler switches. The broader its bandwidth, the lower the CTR is. You might think of CTR as the gain bandwidth product of an electronic amplifier. Photodiodes are very fast switching devices with 10 to 20 nanosecond switching speeds but have CTRs less than 1%. Leakage current is also on the order of a few nanoamps. Phototransistors have CTRs in the 5 to 100 range; however, their switching speed is less than that of a photodiode and have leakages ten times as great as a photodiode. The photodarlington transistor pairs have very high CTRs of 50 to 1,000 so an

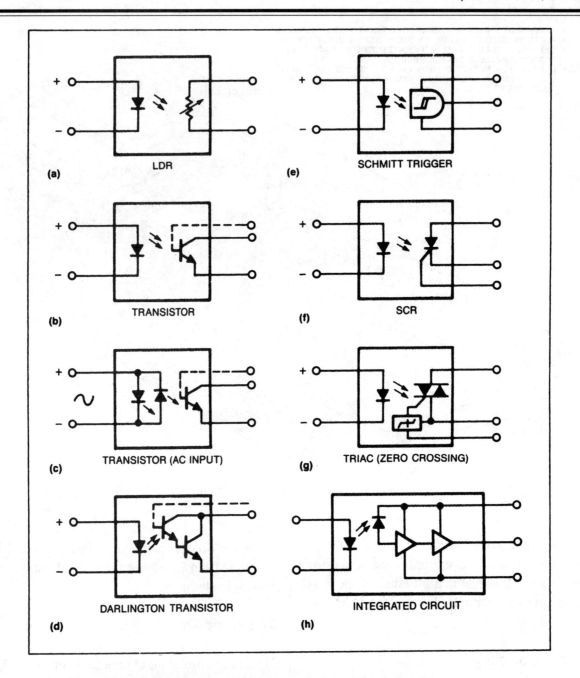

Figure 4-8. The vast array of optocouplers.

input signal can actually be amplified ten fold. The drawback though is slow switching speed and increased leakage current.

An analog application of an optocoupler has its CTR reflecting the measure of its gain or input current to output current. If you intend on using an optocoupler in a digital application, you need to note the CTR at a low collector voltage since many new systems now use 3 volt logic. Also, in digital use, you may see the optocoupler's CTR expressed as its "fan in" or its ability to drive a certain quantity of gates of a specific logic family, such as CMOS. CRT is not a constant for all magnitudes of input current. This is partially due to the Darlington connected transistors on the output having their gains change, especially from temperature. In designing circuits, you'll also need to know that an optocoupler's CTR dwindles in time. The primary culprit is the photon emitter simply

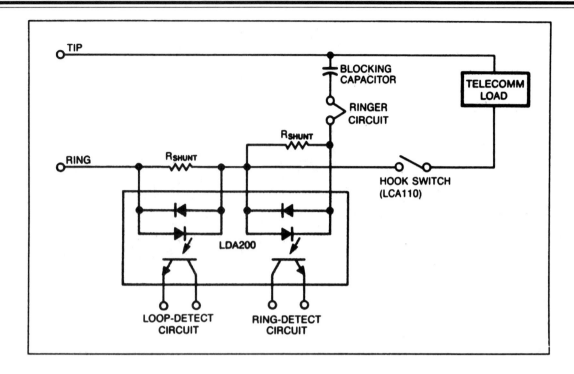

Figure 4-9. A phone line optocoupler application.

wearing down and emitting less photons with increased age — optoelectronic geriatrics! Most optocouplers use the GaAlAs emitter; but, Philips in the Netherlands, and their American subsidiary Amperex, used to reverse the order of these elements (AlGaAs) and use a single heterojunction with two epitaxial layers and different band gaps. Philips claimed this combatted the effects of CTR degradation.

DIP Isolators

Optocouplers come in dual in-line (DIP) and surface mount packages. A typical device has the GaAsP emitter and silicon photodetector attached to wire bonds on separate frames. Then they have a coat of silicone deposited on the junction. There is an insulating transparent film next sandwiched

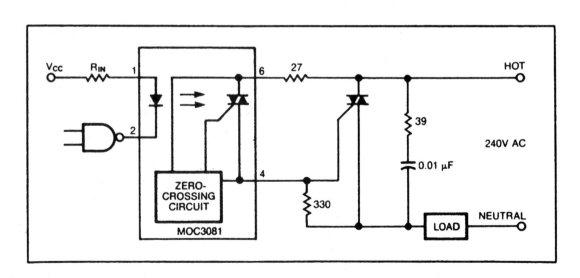

Figure 4-10. A more conventional isolation optocoupler application.

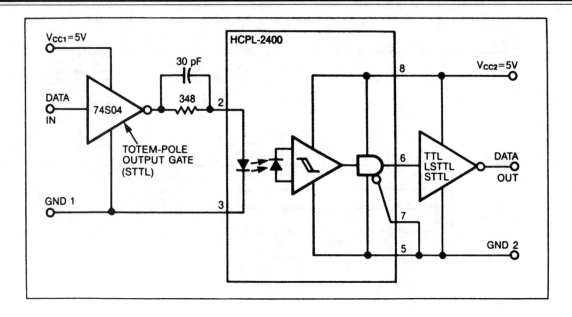

Figure 4-11. A fast switching optocoupler application.

between the units. (See *Figure 4-6*) Last, the entire package is encapsulated with epoxy.

PC Board Layouts

Optocouplers need to have attention directed at minimizing the capacitance between their inputs and outputs. The most effective way to do this is to physically separate the optocoupler's input and output circuitry. (See *Figure 4-7*) If you use the optocoupler as a line coupler, you should properly dress the cable to minimize stray capacitance. You can also place a special ground trace under the optocoupler to serve as a shield since its transistor's base is very susceptible to common mode transients. You may also snip off pin 7 of the IC package in the optocoupler in *Figure 4-4* to reduce any stray input into the lower transistor in this split Darlington configuration.

Since most optocouplers have an open collector output, excessive shunt capacitance there limits both rise and fall times. Therefore, physically place the driven logic gate as close to the optocoupler output as possible. You may also connect an external resistor between pins 5 and 7 to reduce propagation delays. Place external bypass capacitors close to the IC to minimize the possibility of oscillations. In practice, a low inductance 0.01 µF capacitor placed very close to and across pins 5 and 8 (V_{CC} and ground) of the optocoupler is an excellent practice. (See *Figure 4-4*)

Optocoupler Variances

As long as these three elements exist: an IR photoemitter, a transparent separating medium, and a matching IR photodetector, you have an optocoupler. You can take advantage of the many active device semiconductors associated with optoelectronics. (See *Figure 4-8*) The primary differences in these various optocouplers is their outputs.

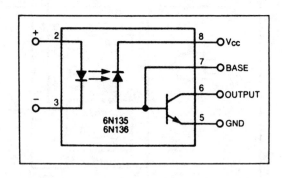

Figure 4-12. The 6N135 generic de facto optocoupler standard.

Chapter 4

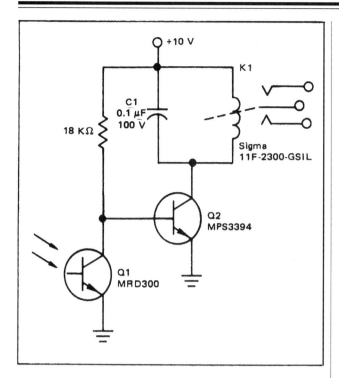

Figure 4-13. An LDR (light dependent resistor) optocoupler application.

Commercially Available Optocouplers

Figure 4-9 shows a Theta-J Corp. LDA200 optocoupler monitoring the phone line to detect the presence of both loop current and a ring signal. The input shunt resistor's value determines the trigger point for the loop current. The LDA200 optocoupler's output is two open-collector transistors rated at 20 volts. *Figure 4-10* shows isolation of high voltages from sensitive input circuits, a more traditional use of an optocoupler. This IC performs on inputs up to 240 volts with an isolation voltage rating of 7,500 volts and an output triac driver which can withstand 800 volts.

The H-P HCPL-2400 is an optocoupler designed for fast switching speeds up to 20 megabaud. (See *Figure 4-11*) The Schottky TTL logic helps attain this speed. *Figure 4-12* shows one of the most popular and generic optocouplers, the 6N135 and 6N136 made by GE, TI, H-P and General Instruments. This is a JEDEC (Joint Electronic Devices) registered and UL safety approved optocoupler.

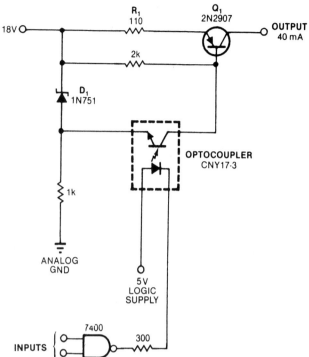

Figure 4-14. A current pulser optocoupler application.

These ICs have a bandwidth of 2 MHz and a switching speed of 1 Megabits/second.

One of the first photodetectors was the LDR (Light Dependent Resistor) which is the basis of *Figure 4-13*, the Clairex CLM51-2, which has a center-tapped LDR. You can connect a 2 kΩ resistor in series or parallel to change the total resistance over a 4:1 range. By varying the IR emitting LED's current from 0.1 to 10 mA you can vary the reference resistance value by 10:1. The LED's maximum current is 40 mA. *Figure 4-14* is a current pulser application using a simple optocoupler with its output pulled low by a 2-input 7400 NAND TTL gate. The constant current feature of this pulser ensures linear rise and fall times and helps eliminate output waveform overshoot. *Figure 4-15* addresses the perennial problem of an optocoupler's slow switching speed. The addition of these discrete components provides negative feedback to the base of the light sensitive transistor, increasing switching speed ten fold. The switching action of the

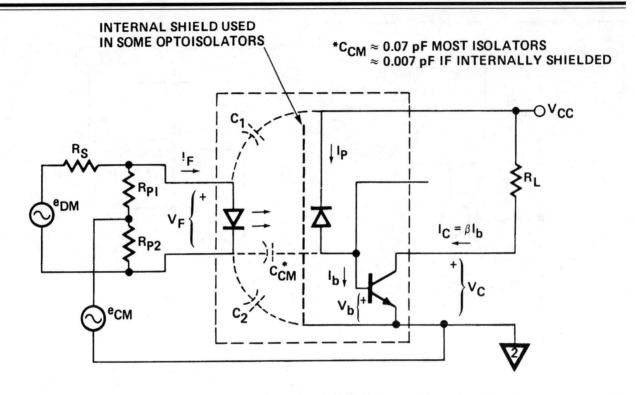

Figure 4-15. Using negative feedback to increase optocoupler speed.

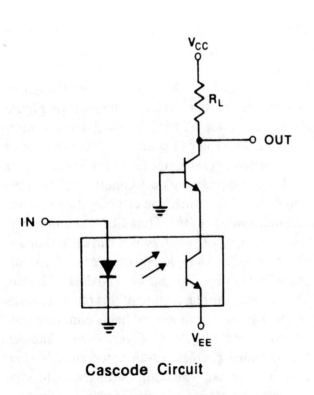

Cascode Circuit

Figure 4-16. Using a cascode transistor to increase optocoupler speed.

external PNP transistor provides Schmitt trigger action on slow ramp input signals.

There are other ways to increase switching speed. *Figure 4-16* is a cascode circuit in which the phototransistor is the current source and the load resistance is collectively composed of isolation transistor's emitter resistance, which you can limit to a few ohms. The key is to select a transistor with a low C_{ob} which will handle enough current to provide crisp switching waveforms. *Figure 4-17* uses an op amp to accomplish the same thing, to lower the effective load resistance by driving the summing point (inverting input) of the op amp by the constant current sourcing of the optocoupler's phototransistor. This point must be at ground level so the phototransistor's collector is at virtual ground. That is why you need a V_{EE} negative supply.

Chapter 4

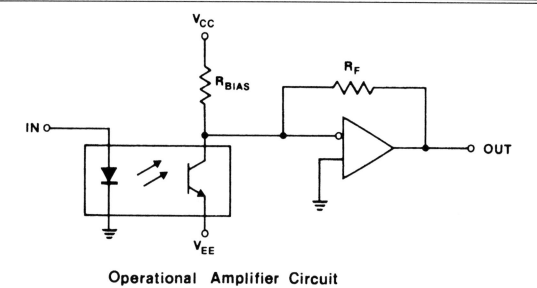

Operational Amplifier Circuit

Figure 4-17. Using an op amp to reduce load resistance and to increase optocoupler speed.

Chapter 4 Quiz

1. An optocoupler:
   A. Provides isolation of grounds from separate circuits.
   B. Couples energy from one circuit to another.
   C. Works like a relay, without the relay's drawbacks.
   D. All of the above.
2. An optocoupler consists of an IR photoemitter, a transparent medium and:
   A. An IR photodetector.
   B. A case with a slit to allow light to enter.
   C. A photometric LED detector.
   D. None of the above.
3. A photoemitter's:
   A. Main concern is coupling efficiency.
   B Voltage drop is more important than gain and bandwidth.
   C. Light emitting area is as large as possible.
   D. B and C.
4. You can use _____ for a photodetector.
   A. A photodiode which is inherently fast.
   B. A phototransistor with the base-collector junction as the light detector.
   C. A photodiode and a separate transistor to amplify the photodiode's current.
   D. All of the above.
5. An optocoupler's response time is:
   A. Inherently slow.
   B. Inherently fast.
   C. Inherently slow due to excessive junction capacitance.
   D. A and C.
6. The most important parameter on an optocoupler's data sheet is:
   A. Insulation.
   B. Isolation.
   C. CTR and response time.
   D. None of the above.
7. Which is true?
   A. A photodiode is the fastest switching device but has the greatest leakage.
   B. A phototransistor is the fastest switching device with very little leakage.
   C. A photodiode is the fastest switching device with only nanoamps of leakage.
   D. A photo-Darlington transistor has a low CTR.
8. Properly dressed cables going to a PC board with an optocoupler on it:
   A. Ensure the signal is impedance matched.
   B. Minimize stray capacitance.
   C. Shield the optocoupler from ambient light.
   D. None of the above.

9. Many optocouplers have:
    A. Open collectors contributing to poor rise and fall times.
    B. Means to reduce propagation delay by inserting an external resistor between two IC pins.
    C. The ability to reduce oscillations if a bypass capacitor is placed physically close to the optocoupler.
    D. All of the above.
10. The 6N135 is:
    A. A de facto industry standard optocoupler.
    B. Made by several manufacturers.
    C. Designed for speed.
    D. A and B.

Chapter 5
Phototransistors and Optointerrupters

Chapter 5
Phototransistors and Optointerrupters

This chapter predominantly covers applications of phototransistors and optointerrupters. They virtually have the same optical and electrical characteristics. Phototransistors are essentially photodiodes with a base. Their main difference is the phototransistor's greater sensitivities or increased gains.

What to Do With a Phototransistor's Base

A phototransistor's third base lead infrequently comes to the outside world. But when it does, you can use it as a transistor and switch the phototransistor on or off through this accessible base. You can still enhance switching, though, without an accessible phototransistor base, as we'll show. *Figure 5-1* adds a base resistor, R_B, to the phototransistor's base and reduces the switching turn-off time. If the phototransistor's base is not available, you can use an op amp buffer (See *Figure 5-2*) to increase switching speeds. *Figure 5-3* is an alternative to *Figure 5-2* and uses an isolation transistor in the phototransistor's collector to increase switching speeds. *Figure 5-4* provides both optimized speed and output voltage for a phototransistor. Transistor Q2's common base stage presents a low impedance to the phototransistor which maximizes speed. Since Q2 has a near unity gain, the current in the load resistor, R_L, approximately equals the phototransistor's current. Therefore, the effect Q2 provides to the phototransistor's frequency response makes it virtually independent of its load. *Figure 5-5* increases switching speeds with an alternative transistor output driver. This phototransistor has two unique features. First, it uses a neon lamp in an AC line voltage presence indicator circuit so the phototransistor's base must be sensitive to the neon lamp's wavelength. Secondly, capacitor C helps stabilize the phototransistor's output.

Photodetector Similarities

You can use a phototransistor and photodiode as interchangeable parts if you use the phototransistor in its photodiode mode. Your optoelectronics application's photocurrent though primarily determines if you need to use:

1. A phototransistor.
2. A photodiode.

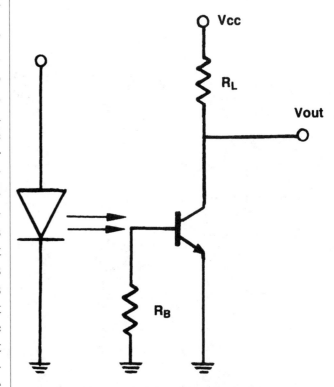

Figure 5-1. Enhancing a phototransistor's switching speed by adding a base resistor, R_B.

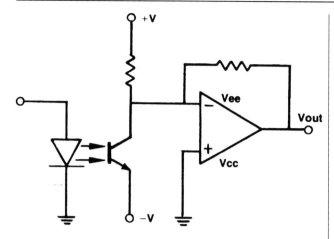

Figure 5-2. Enhancing a phototransistor's switching speed by using an op amp buffer.

3. An avalanche photodiode.
4. A photodiode with an external amplifier.

Your application's speed of response and available light also determine photodetector type selection.

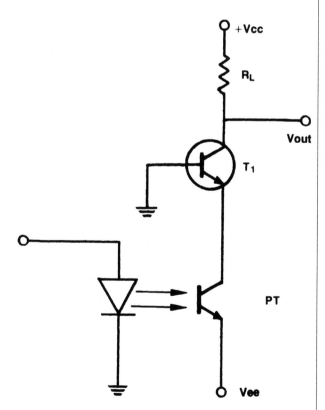

Figure 5-3. Enhancing a phototransistor's switching speed by using an isolation transistor in the phototransistor's collector.

The Photon Effect in Semiconductors

Let's quickly overview photon generation in semiconductors and then specifically apply this to phototransistors. Light with the proper wavelength and intensity casting upon a semiconductor crystal increases the concentration of charge carriers. (See *Figure 5-6*) This semiconductor band structure has a vitally important energy gap, called the band gap, or forbidden region (E_g). To demonstrate its sensitivity and wavelength dependence on light, let's use two different light sources, f_1 and f_2, with photon energies of hf_1 and hf_2 respectively. (See *Figure 5-6*) Photon energy hf_1 of f_1 adequately passes beyond the band gap energy level. Conversely, photon energy hf_2 of light f_2 just barely reaches the band gap. It is inadequate to penetrate the band gap and reach the next region, the conduction band. Light, such as f_1, which is energetic enough to penetrate the band gap transfers its energy to site one in the valance band. This excited electron attains a higher energy level, liberating it to serve as a current carrier. The less energetic light source, f_2, barely attains the band gap and its depleted energy causes it to fall back into the valence band and recombine with a hole at that site.

The momentum and density of hole-electron pairs is greatest at the valence and conduction bands' centers. This is clearly an energy dependent response. A less abstract or obtuse example (See *Figure 5-7*), is a cadmium-selenide crystal, a material from which photocells are made. From our previous discussions, it seems probable to expect

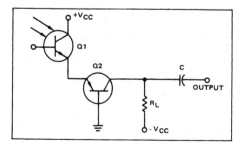

Figure 5-4. Enhancing a phototransistor's switching speed and output voltage with a nearly unity gain transistor stage.

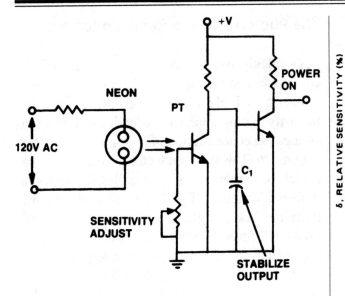

Figure 5-5. A phototransistor with a neon lamp input.

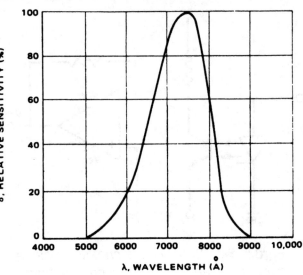

Figure 5-7. A cadmium selenide crystal's sensitivity curve.

symmetry in this curve. However, it slopes more steeply to the right. Trapping centers and other absorption effects cause this. Adding impurities to bulk semiconductor material also changes this intensity-to-wavelength curve. Lastly, photo-excitation in the conduction band also shifts this response curve.

In our first applications we use a series of Motorola MRD series phototransistors. Their spectral sensi-

tivity curve is not perfect (See *Figure 5-8*), but much more closely resembles a bell-shaped curve. Lenses can help increase irradiance on a photodetector (See *Figure 5-9*); But you must always be vigilant to spectrally match the radiation sensitivity of a photodetector to a photoemitter's wavelength, as well as be keenly aware of the emitter's radiation pattern.

The P-N Junction's Photo Effect

Predominant current flows across the P-N junction under proper light exposure and reverse bias-

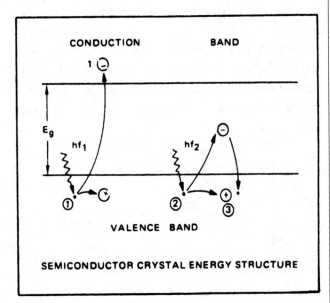

Figure 5-6. Light increasing charge carrier concentrations.

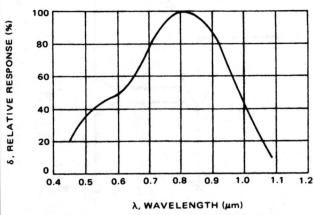

Figure 5-8. A typical phototransistor's less skewed sensitivity curve.

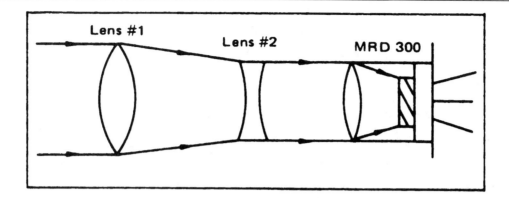

Figure 5-9. Lenses helping to concentrate light onto a phototransistor's base.

ing. (See *Figure 5-10*) Photons create hole-electron pairs on both sides of the P-N junction. A phototransistor uses this P-N junction for its collector-base diode of a bipolar transistor. This photo induced current is the transistor's base current, the smallest of its three currents. This base may be left omitted, open or internally tied to a quiescent level. In the phototransistor's energy diagram, *e*, the applied electric field, causes the collector to pull in electrons supplied to the emitter by the base. This junction's photon driven electrons conduct from the P-side to the N-side and then into external circuitry. Thus, it is merely a photodiode.

Color Temperature Sensitivity

A phototransistor is color temperature sensitive. Since ambient light sources typically have such broad ranges and varied spectral contents, manufacturers often use glass or plastic filters over their exposed light gathering bases. Glass affects other phototransistor optical characteristics. *Figure 5-11* shows a graph in which the narrow plot corresponds to the inner lens and the broader response is from a flat glass lens. The phototransistor's window size also affects its response curve.

Phototransistor Applications

Figure 5-12 is a simple circuit using a phototransistor which responds to the light's presence by activating relay K1. Transistor Q2's gain is approximately 55 and a base current of about 0.5 mA saturates this transistor. Phototransistor Q1 adequately produces this drive and has a sensitivity of 4 µA/foot candles. With enough light, it drives Q2 into conduction to activate relay K1; how much light is "enough"? The following equation

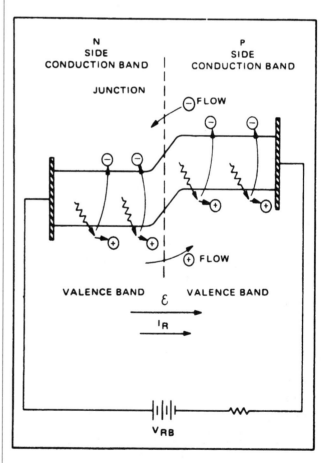

Figure 5-10. A reversed biased P-N junction reacting to light.

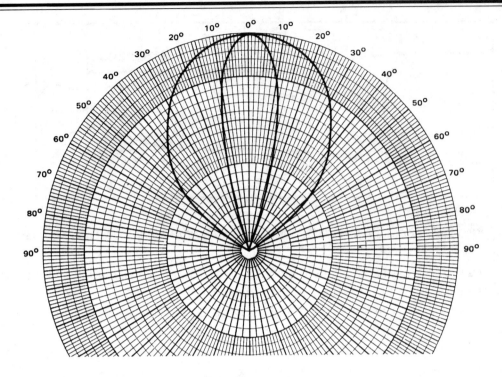

Figure 5-11. *A phototransistor's lens vs. a flat glass base cover's sensitivity patterns. (Courtesy of Motorola)*

describes this threshold of light to energize relay K1 as:

$$\text{Light}^{THRES} = I_{BQ2}/Q_{BQ1SENS}$$
$$= 0.5 \text{ mA}/4 \times 10^{-6}/\text{fc} = 125 \text{ fc}$$

Or 1346 Lamberts/m², in more modern terms.

As you will recall, the reverse saturation current which flows in the reverse biased collector-to-base junction with the emitter open circuited is a phototransistor's dark current, or I_{CBO}. The phototransistor amplifies dark current in the collector and this is designated, I_{CEO}, referring to collector-to-emitter current with the base open circuited. Since it is very common for a phototransistor to have no base or to operate with an open circuited base, the parameters relating incident radiation in the collector-to-base junction to the base-to-collector photocurrents are crucial. Many data sheets on phototransistors therefore show both photodiode and phototransistor operating curves for the phototransistor. These parameters of photocurrent versus incident radiation are the collector-to-emitter sensitivity S_{RCEO} and the collector-to-base radiation sensitivity, S_{RCBO}.

The phototransistor's gain (b) is this ratio.

$$\beta = S_{RCEO}/S_{RCBO}$$

The phototransistor in *Figure 5-13* conducts and drives Q2's base causing Q2 to conduct, pulling its collector low. This provides a path for relay K1. When phototransistor Q1 conducts in the presence of light, its collector and Q2's base are approximately 0.3 volts. Light's presence, in this case, de-energizes relay K1.

Figure 5-14 further expands on our light activation circuit to activate an SCR (Silicon Controller Rectifier) alarm system. Interrupting Q1's light energizes the relay and provides the SCR with triggering current to latch ON. Only breaking the light beam activates this alarm, and momentary contact switch S1 disarms it. You can further enhance this circuit by using a sensitive gate SCR and eliminate the need for a relay. (See *Figure 5-15*) The

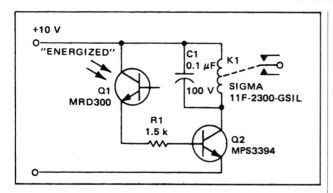

Figure 5-12. A phototransistor light presence detector energizing a relay. (Courtesy of Motorola)

phototransistor holds the gate low, as long as light persists. Interrupting the beam pulls the gate to a triggering level. S1 is again a reset switch.

Figure 5-16 is a voltage regulator for an incandescent lamp projector. An incandescent bulb's light output and its color temperature over a limited range (See *Table 2-7*) are *rms* voltage dependent. Maintaining a constant light intensity under AC line variations requires an AC voltage regulator.

Figure 5-16 provides an unusual feedback arrangement through a current diverting mechanism. Pot R6's position, the magnitude of charging current and the 0.1 µF capacitor, C, all influence the firing angle of the UJT (Unijunction Transistor), B1. The UJT, and the PUT (Programmable Unijunction Transistor), like SCRs and triacs are members of a broad semiconductor group called thyristors. They exhibit many unique features, the most salient of which is their latching effect. Once a control current turns them on, they don't require any further control current to sustain or keep them on. A thyristor is a four-layered diode with an anode, cathode, and a gate. The gate accepts pulses and allows conduction in just one direction. Conduction continues until you remove the anode or cathode voltage.

The projector's lamp brightness sets the current in Q3, the phototransistor. The greater the phototransistor's current, the more current it diverts from the UJT's 0.1 µF timing capacitor. Pot R6 sets the desired brightness level as well, after

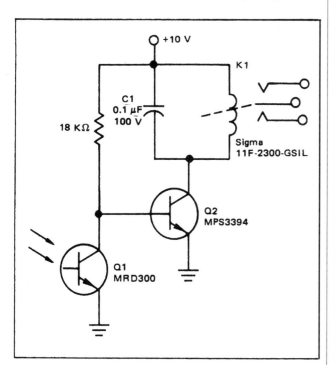

Figure 5-13. A phototransistor detector de-energizing a relay by light's presence. (Courtesy of Motorola)

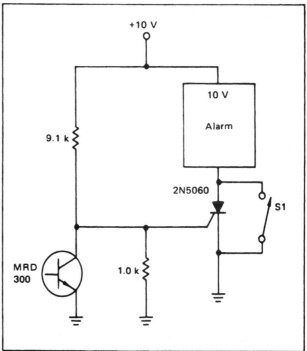

Figure 5-14. A phototransistor driving an SCR alarm circuit. (Courtesy of Motorola)

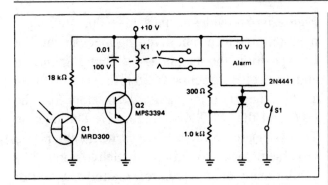

Figure 5-15. Using a sensitive gate SCR to eliminate the relay. (Courtesy of Motorola)

the phototransistor helps regulate it against AC line variations.

Optointerrupters

These are strategically placed, or opposite each other, light emitters and detectors with optically matched wavelength sensitivities. Most commonly a plastic molding houses them with a small slot separating them. This intentional slot (See *Figure 5-17*) provides access or a path for a moving, typically rotating, object to break the light beam. One of the first and most pervasive applications of this technology was the old IBM punched cards. Today many cars use optointerrupters to sense both the speed and direction of rotating parts.

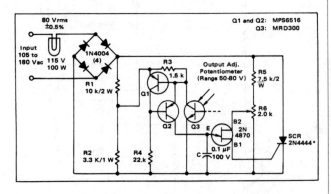

Figure 5-16. A phototransistor based AC line voltage regulator to ensure uniform intensity in a projector's incandescent bulb. (Courtesy of Motorola)

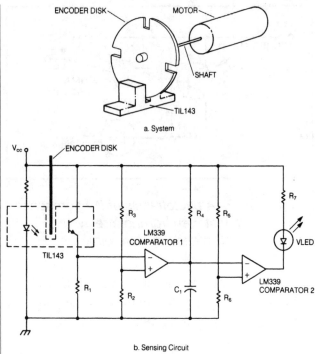

Figure 5-17. A slotted optointerrupter circuit which senses rotational speed and drives a low speed LED indicator alarm. (Courtesy of Texas Instruments)

Figure 5-17 is an application of speed sensing using an optointerrupter. This circuit lights a warning light if the shaft's speed falls below a certain acceptable level. The encoder or four notched wheel allows light to pass four times a revolution. When the slot is present the optointerrupter has unimpeded light and therefore more gain. This causes a voltage increase at comparator 1's (-) inverting input. Resistors R2 and R3 form a voltage divider, setting a voltage reference on the non-inverting input (+) of the quad op amp comparator LM339. The breaking beam causes comparator 1's output to be a square wave with a limited duty cycle which is the ratio of open to blocked area on the rotating encoder disk. You can select the RC values of R4 and C1 so C1 discharges and then charges, but never charges above the reference voltage established by voltage divider R5 and R6.

If the speed **does** slow and allows C1 sufficient charging time, its amplitude exceeds the reference voltage on the non-inverting (+) input of compara-

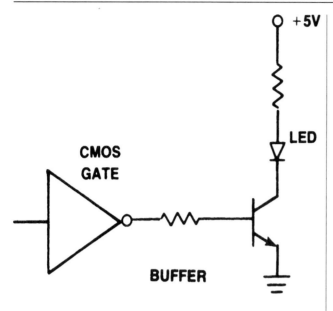

Figure 5-18. *A double inverting CMOS current buffer.*

tional speed. The LM339 is an open collector output op amp with more than a normal op amp's output current sinking capability. If you use an ordinary op amp you may have to buffer the output with a non-inverting buffer (current amplifier) to maintain this circuit's convention of a lit LED signifying a slow rotation speed. *Figure 5-18* uses a CMOS inverter and transistor to double invert the signal, maintaining its logic state. Some non-inverting CMOS buffers alone will drive more efficient LEDs.

Figure 5-19 is another optointerrupter circuit which uses a dual timer. The timer's first half (to the left) is an astable multivibrator or a free running squarewave oscillator which creates pulses. These trigger the optointerrupter LED's source (drawn with dashed lines around it). This periodically resets the timer's other half (a missing pulse detector). Interrupting the light beam cuts off these periodic reset pulses and generates an alarm if no more pulses occur with in an interval exceeding

tor 2. This forces comparator 2's output low which sinks current, turning on the visible LED (VLED). The LED indicates a condition of too slow a rota-

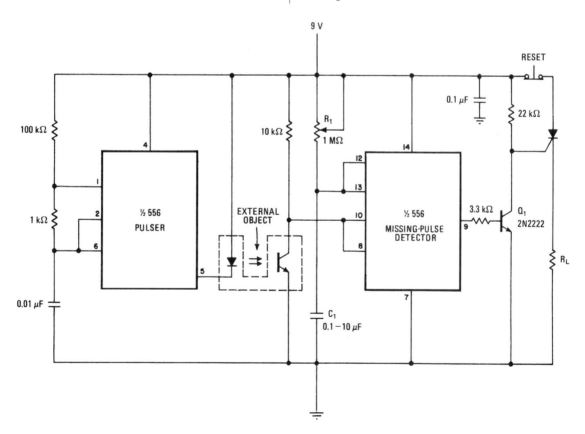

Figure 5-19. *An optointerrupter based on a dual timer.*

the preset period you determine. The reset switch allows you to reset this circuit.

Interesting Phototransistor Based Construction Projects

These three journal articles each build a phototransistor based project. These all relate to hobbies of:

1. Model railroading.
2. Telescope making.
3. Your PC.

The first article, written by John S. Atkinson, "Fun with Phototransistors for Model Railroading," appeared in the August, 1994 issue of *Model Railroad*, page 102. The second article, written by Flodqvist Cote, appeared in the Oct. 1995 issue of *Sky and Telescope* on page 85; "Detecting the Polar Lights." James J. Barbarello wrote the third article, "Card Reader for Your PC," which appears in the August, 1995 issue of *Electronics Now*, page 63.

Chapter 5 Quiz

1. A phototransistor is very different optically and electrically from a photodiode, T or F?
2. A phototransistor usually has its base omitted, T or F?
3. A phototransistor can be operated in a photodiode mode, T or F?
4. Light increases the concentration of charge carriers in a semiconductor crystal, T or F?
5. What factors also determine which type photodetector you select?
    A. Speed of response.
    B. Available light.
    C. Both A and B.
    D. None of the above.
6. What factors contribute to a cadmium-selenide crystal's unsymmetrical response curve?
    A. Absorption effects such as trapping centers.
    B. Adding impurities to bulk semiconductor material.
    C. Photoexcitation in the conduction bands.
    D. All the above.
7. A UJT is also what?
    A. A self oscillating semiconductor.
    B. A thyristor.
    C. A light sensor.
    D. A light detector.
8. The light projector's AC voltage regulator uses:
    A. A current diverting mechanism.
    B. A triple feedback loop.
    C. No means of feedback.
    D. None of the above.
9. An incandescent bulb's voltage, over a limited range, determines its color temperature, T or F?
10. An optointerrupter typically:
    A. Detects light beam interruptions.
    B. Detects rotating movement.
    C. Uses a wavelength sensitivity matched optodetector and optoemitter pair.
    D. All the above.

Chapter 6
Optical Triac Drivers

Chapter 6
Optical Triac Drivers

This chapter investigates the theory of operation and applications of an IR optically coupled triac driver. This is a rather specialized, but very useful, optoelectronics device. The Motorola MOC3011 is our representative industry device. But first, what is this driven device, the triac? It is a solid state latching device within a broad semiconductor family called thyristors. Generally, a thyristor is a four-layered diode with an anode, cathode, and a gate. As Chapter 5 explained, these semiconductors exhibit many unique features, the most important of which is their latching effect. Conduction in a thyristor continues until you remove the thyristor's anode or cathode voltage.

Real World Thyristors

In an SCR (Silicon Controlled Rectifier), a type of AC control and latching thyristor, the gate accepts pulses. This allows conduction in just one direction, as well as control over how much of one half of the AC cycle conduction current flows. It accomplishes this by controlling the point (angle) at which conduction starts. This therefore controls the amount of AC energy under the AC curve. More practically, this translates into how much AC energy is delivered to the SCR's load.

Once a control current turns on an SCR or triac, they require no further sustaining voltage. Don't confuse this with the MOC3011 detector's 100 mA control or "holding" current to sustain or keep them on, as we'll later discuss. You can visualize a triac as two SCRs placed in parallel but opposite directions. This permits conduction and control over both halves of the AC cycle. In contrast, an SCR has control over only one half of the AC cycle's starting conduction angle. Therefore, the triac has greater control of the AC energy delivered to loads attached across the 110 VAC power line.

IR Optically Coupled Triac Drivers

These optoelectronic devices consist of a gallium-arsenide IR LED optically exciting a silicon detector chip. This special design of optodetector optimizes its parameters to specifically drive the gate of a larger triac. Optical coupling in a MOC3011, just like with an optocoupler IC proper, isolates AC loads from the driving circuitry. The MOC3011's optical energy, not its electrical energy, controls the triac's gate to realize this isolation. These two circuits have no common ground; therefore, no common electrical currents flow between them.

The MOC3011's forward IR LED diode has a 1.3 volt drop across it when pulling 10 mA and a reverse breakdown voltage greater than 3 volts. It can pull or sink a maximum of 50 mA. *Figure 6-1* is a greatly simplified schematic of an IR optically coupled triac driver. The Greek letter lambda (λ) in this figure, as is common practice in the industry, represents light radiation.

The MOC3011 IR optically coupled triac driver's detector passes 100 mA in either direction and blocks a minimum of 250 VDC, also in either di-

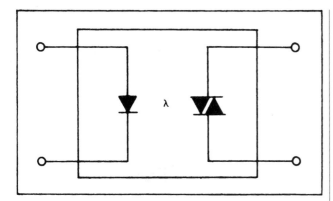

Figure 6-1. A greatly simplified schematic representation of an IR optically coupled triac driver.

rection of current flow. After triggered into its conduction state, this IR optically coupled triac driver remains in that state until it drops below its holding current of about 100 mA. You can trigger the detector by any of the following three methods:

1. Exceeding the forward blocking voltage.
2. Voltage ramps across the detector which exceed its static dv/dt rating (to be explained).
3. By photons from the IR LED.

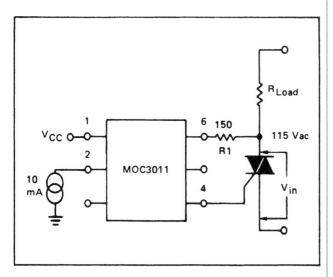

Figure 6-2. The absolutely simplest triac drive method using a minimum component count.

Driving Actual Loads

Figure 6-2 shows the MOC3011 in a very simple triac gating circuit. Note the pin numbers of this 6-pin mini-DIP device, and that pins 3 and 5 have no connections. This circuit is satisfactory for driving mostly purely resistive loads. However, most loads across an AC source are motors, solenoids, etc., which are inductive loads. These cause several problems. Primarily, inductive loads have their AC current and voltage out of phase. This presents problems for both the triac and the MOC3011 since the triac turns off at zero current while its applied voltage is high. Therefore, this appears to the triac as a sudden rise in applied voltage which turns it on at a rate exceeding its commutating dv/dt rate and the static dv/dt rate of the MOC3011.

Solving Inrush Current

If you use a snubber (See *Figure 6-3*), you limit an inductive load's inrush current or high surge currents upon AC turn-on. This reduces the rate of voltage change the triac experiences at turn-on. An example of unexpected inrush surges occurs when an incandescent lamp bulb first turns on. See the Burglar Baffler in Chapter 10 for a detailed explanation of two circuits, a zero crossing detector and a snubber, to combat this. Think of what a transient surge is. It is a high voltage occurring within a small amount of time. This dv/dt ratio is what is so damaging to electronic circuitry.

The snubber circuit in *Figure 6-3* has a capacitor, C1, and an inductor, Z_L, and two resistors, R1 and R2. Collectively, they protect against sudden inrush currents at AC turn-on. You will still have to select a triac which will withstand this inrush current. But the snubber helps, and in drastic cases, you can even use a snubber on both the MOC3011 and the triac. Ideally, you benefit greatly by knowing the load's power factor. Power factor is the cosine of the load induced phase shift. But since you may not always know this or be able to measure this, you can make a close approximation using a

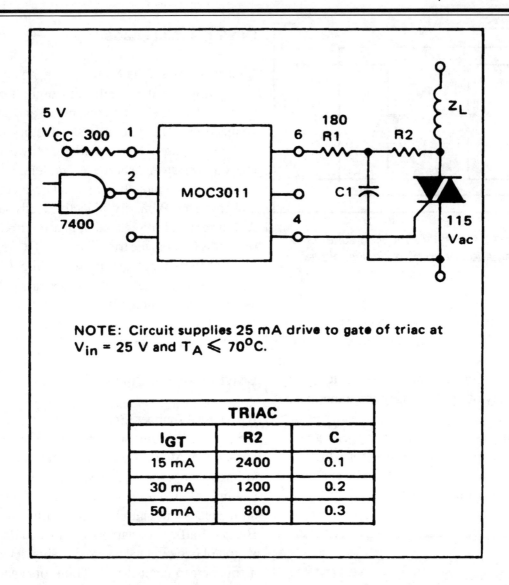

Figure 6-3. A "snubber" circuit to drive inductive loads while suppressing "inrush" AC turn-on surges and transients. (Courtesy of Motorola)

typical power factor. Assuming you have an inductive load with a power factor of 0.1. The triac tries to turn off when the applied voltage is:

$$V_{TO} = V_{pk} \sin \phi \sim V_{pk} \sim 180 \text{ volts}$$

Where V_{TO} is voltage at turn on and V_{pk} is the peak voltage. Let's also assume we want to limit the current through the snubber to 1.2 Amps. Therefore:

$$R1 = V_{pk} / I_{MAX} = 180 \text{ volts} / 1.2 \text{ Amps} = 150 \Omega$$

You will next have to set the time constant $(\tau) = R2 \times C$. Further assuming the triac turns off quickly, the peak rate of dv/dt rise in the MOC3011 is:

$$dv/dt = V_{TO} / \tau = V_{TO} / R2 \times C$$

Using the graph in *Figure 6-4* to determine the temperature at 70 °C we find dv/dt = 0.8 volts/μsec = 8×10^5; therefore:

$$R2 \times C = V_{TO} / (dv/dt) = 180 / (8 \cdot 10^5) \sim 230 \times 10^{-6}$$

118

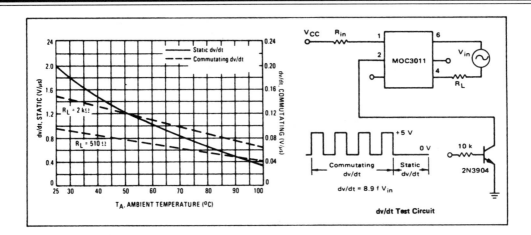

Figure 6-4. The static and commutating dv/dt principle. (Courtesy of Motorola)

The previous equation states the R2 x C product has to be this; therefore, if we allow R2 to be 2.4 kΩ and C to be a 0.1 µF capacitor, we have met our snubber requirements. But there is one last consideration. You can use a sensitive gate triac, such as we did in Chapter 5's alarm circuit and delete a relay. The MOC3010 is virtually an identical IC to the MOC3011; however, it requires 15 mA of current for triggering. The MOC3010 sensitive gate IR optically coupled triac driver uses just 15 mA and our calculations were based on this device. This involves using a lower value of R2 and a correspondingly higher value of C, see the small table within *Figure 6-3*. Using a triac gate one half as sensitive, a 30 mA device, halves R2 and doubles C. A triac gate approximately one third as sensitive (50 mA) cuts the 2.4 kΩ R2 value in thirds, just 800 Ω, but correspondingly triples the value of C to 0.3 µF. The R2 x C product will always remain constant.

The dv/dt Concept

Figure 6-4 shows an ambient temperature related graph and waveform of both static and commutating dv/dt. Note the scale for each dv/dt is volts/µsec. That shows how severe these spikes can be. The triac experiences commutating dv/dt while the MOC3011 IR optically coupled triac driver experiences static dv/dt. Occasionally, transients on the

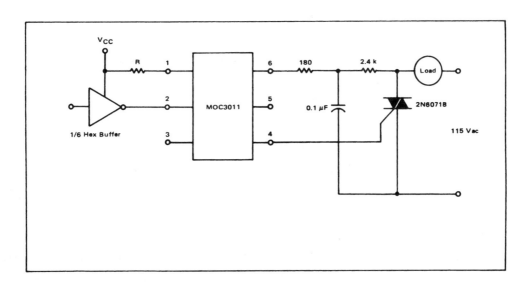

Figure 6-5. A CMOS to load triac driver interface. (Courtesy of Motorola)

The MOC3011's Input Circuit

Figure 6-5 shows a simple CMOS hex buffer interfacing to the MOC3011's IR LED (pin 1) with just a series resistor, R. The resistor's value should take cognizance of the MOC3011 LED's limit of a 10 mA to 50 mA range. A value of 15 mA is ideal since currents in excess of this do not increase speed. They just shorten the MOC3011's life due to prematurely degrading the CTR of this device's internal opto-coupling mechanism. See Chapter 4 for a detailed description of this.

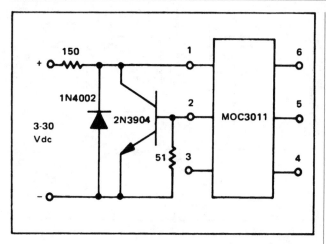

Figure 6-6. *An input protection circuit for the MOC3011 IR optically coupled triac driver. (Courtesy of Motorola)*

You can further protect the MOC3011's input circuitry by using the circuit in *Figure 6-6*. In applications such as solid state relays (to be covered), the input voltage varies widely and therefore so does the MOC3011's input current. This circuit allows an appropriate but more limited input voltage range to drive the MOC3011's IR LED. Nonetheless, it maintains the MOC3011's forward current at about 15 mA and also guards against an inadvertent application of reverse polarity.

AC line exceed the MOC3011's static dv/dt or rate of voltage change divided by the rate of time during this change. When this occurs, it is usually not a problem because the triac driver circuitry triggers at the zero crossing current. The MOC3011's zero crossing circuitry is responsible for forcing the triac to turn off again at the next half cycle of AC.

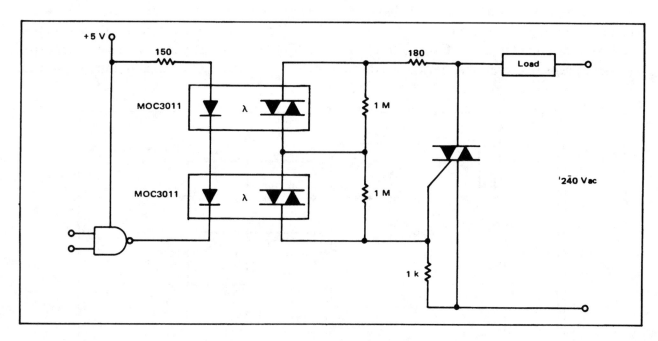

Figure 6-7. *Connecting two MOC3011 IR optically coupled triac drivers in series to accommodate 240 VAC. (Courtesy of Motorola)*

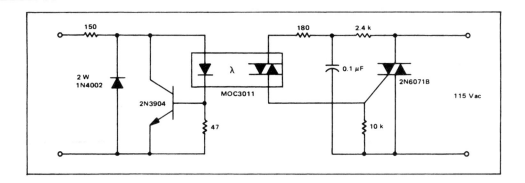

Figure 6-8. A solid state relay using an MOC3011. (Courtesy of Motorola)

Applications of The MOC3011

The MOC3011's parameters do not allow it to drive 240 VAC. However, you can connect two MOC3011s in series to realize this goal. (See *Figure 6-7*) This circuit ensures an equal voltage division across the two IR optically coupled triac drivers by making the two resistors 1 MΩ, each equally divide the voltage in half. The high values of these resistors also ensure they do not draw or divert very much current.

Figure 6-8 is a solid state relay based on an MOC3011 IR optically coupled triac driver IC. This circuit has protection for its input just as we used in *Figure 6-6* and a snubber on its output which drives the 15 mA sensitive gate triac circuit in *Figure 6-3*. Note the addition though of the 10 kΩ resistor. The Burglar Baffler construction project in Chapter 10 also places two IR optically coupled triac drivers in series with just 110 VAC across them. This greatly increases their life due to minimal stress and even more greatly reduces their CTR degradation effects.

Figure 6-9 is an application of remotely controlling and driving a triac. This application meets most city and local building code wiring requirements to have all 115 VAC light switch wiring contained within a conduit or pipe. By using a triac and a low voltage source of just 5 volts, you can control large lighting loads from a long distance away. Such low voltage wiring of just 5 volts does not usually require its housing within a conduit. This is a great savings to the builder or building maintenance personnel. The circuit we use could also drive an attic fan, swimming pool motor or pump.

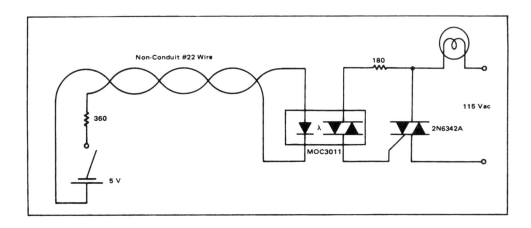

Figure 6-9. Remotely controlling AC loads and lighting to avoid enclosing them within conduit to meet building codes. (Courtesy of Motorola)

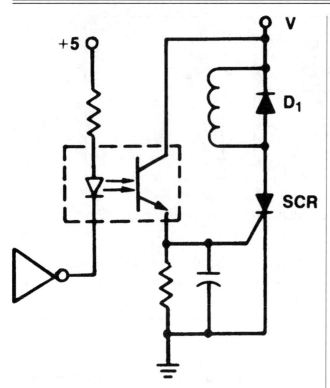

Figure 6-10. A basic optocoupler SCR drive scheme.

Comparing Thyristor Drive Schemes Using Optocouplers

Figure 6-10 shows the basic principle of operation with an SCR type thyristor in this case. When the inverter's output goes low, the LED within the

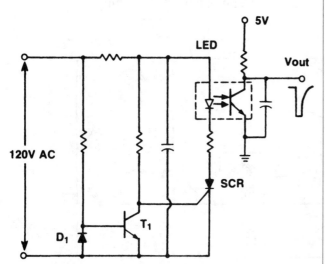

Figure 6-11. A diode half-way rectifier circuit which realizes zero crossing SCR activation.

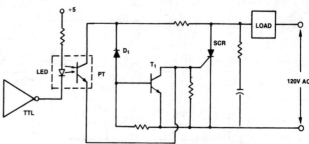

Figure 6-12. A 60 Hz synchronizing circuit to turn on its loads simultaneously or at the same spot on the AC waveform.

optocoupler fires or activates the SCR's gate. The SCR conducts and drives its inductive load. The RC input circuit guards against false triggering. Diode D1 across the load prevents spikes. This circuit is very elementary compared to our MOC3011 drive circuits.

Figure 6-11 is a zero crossing detector to turn on an inductive load right as it passes through zero. The diode D1 is a half wave rectifier to control the transistor. The transistor supplies a pulse to the SCR's gate. This further produces a negative going pulse at the optocoupler's transistor. This allows synchronization of other circuits. This zero crossing detector's actions also reduce RFI or radio frequency interference.

Figure 6-12 is again a circuit which is triggered by the inverter and uses the 60 Hz synchronizer in *Figure 6-11* to fire the SCR in the cross over re-

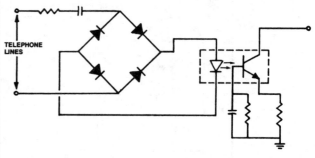

Figure 6-13. A telephone ringer circuit which triggers a triac.

gion to start the load. This is a zero crossing detector which the MOC3011 has already built in it.

Figure 6-13 is a simple telephone ringer circuit with an extra transistor for increased gain. It triggers a triac to drive a heavy load. So you could use this circuit to turn on AC devices or light appliances while away from your house if you could guarantee all other calls would be suppressed.

Chapter 6 Quiz

1. An IR optically coupled triac driver is a member of what semiconductor family?
   A. Regenerative detectors.
   B. Thyristors.
   C. Pinch-off devices.
   D. None of the above.
2. An IR optically coupled triac driver's most important feature is its:
   A. Ability to withstand 250 °C.
   B. Ability to change the phase relationship of AC current and voltage in its inductive loads.
   C. Latching effect.
   D. All the above.
3. A thyristor, such as an SCR or triac, controls the angle at which the AC cycle begins conduction and this, in turn, controls:
   A. The phase relationship between the IR optically coupled triac driver and the triac.
   B. The amount of energy delivered to the load.
   C. Both A and B.
   D. None of the above.
4. How do SCRs and triacs differ?
   A. Triacs control both halves of the AC cycle.
   B. SCRs control both halves of the AC cycle.
   C. Both A and B.
   D. None of the above.
5. The IR LED in optically coupled triac drivers is made of:
   A. Silicon.
   B. Selenium composites.
   C. Gallium arsenide.
   D. None of the above.
6. An IR optically coupled triac driver realizes isolation because:
   A. The triac and it use no common ground.
   B. The triac and it experience no common current flow.
   C. A and B.
   D. None of the above.
7. The Greek letter l signifies:
   A. Light radiation.
   B. The absence of light.
   C. A warning of excessive voltages.
   D. None of the above.
8. You can trigger the IR optically coupled triac driver's detector by photon flow, T or F?
9. The MOC3011 has no connections to pins 1 and 6, T or F?
10. Usually, inductive loads have their voltage and currents in phase, T or F?
11. Turning on an incandescent light bulb can result in an AC inrush current, T or F?
12. It is not very beneficial to know a load's power factor, T or F?

13. The MOC3010 IR optically coupled triac driver IC uses just 15 mA; therefore, you use it with sensitive gate triacs, T or F?
14. The snubber circuit for a less sensitive gate triac uses a larger resistor and a proportionately smaller capacitor, T or F?
15. A value of 15 mA is ideal for the MOC3011 IR optically coupled triac driver IC's IR LED, but increasing it produces greater speed and a longer life by reducing its CTR degradation, T or F?
16. The protection circuit in *Figure 6-6* guards against inadvertent reverse polarities, T or F?
17. The 1 MΩ resistors in *Figure 6-7* equally divide the IR optically coupled triac driver's AC by the volt-age division principle, T or F?
18. The circuit in *Figure 6-9* uses AC as a control source over the lines which control the MOC3011's input, T or F?

Chapter 7
Photoelectric Sensing

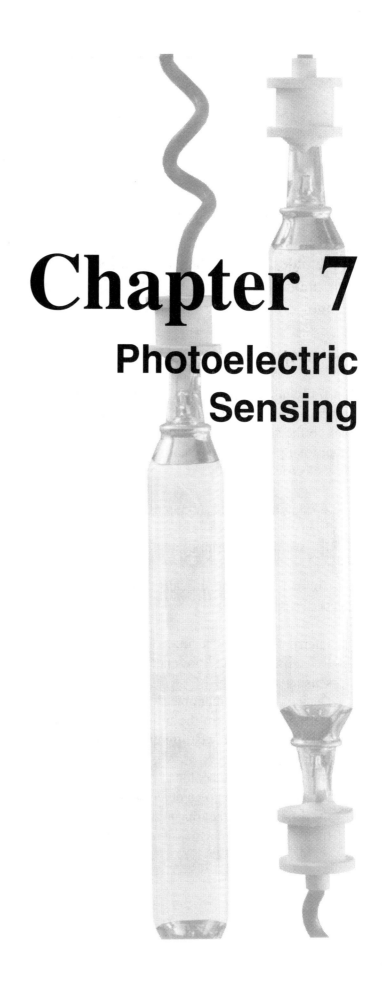

Chapter 7
Photoelectric Sensing

This chapter examines the theory and applications of photoelectric sensing. We're all familiar with the IR photoelectric sensor which opens a grocery store's door. This is one of the oldest rudimentary applications of photoelectric sensing technology. More elaborate photoelectric sensing schemes and sensors have evolved since this 40 year old technology's infancy, such as lasers, which we cover.

Photoelectric Sensing Modes

Generally, photoelectric sensing detects a target with an optoelectronic device which either reflects back to the sensor, is interrupted, or its signal is either totally or partially absorbed. *Table 7-1* lists photoelectric sensors by:

1. Sensing modes and submodes.
2. Type of light source used.
3. Type of output circuit used.

The following more specifically describes photoelectric sensing modes:

1. Interrupting a light beam.
2. Detecting an object's presence anywhere in the sensing area.
3. Detecting transparent objects.
4. Detecting an object by the difference of the light intensity between two different receiving photoelectric sensors.
5. Detecting a target's color.
6. Projecting light onto an object and detecting the difference in the transmitted and received light's angles of transmission and arrival.

However, some manufacturer's literature states photoelectric sensing modes differently as:

1. The opposed mode.
2. The retroflective mode.
3. The proximity mode.

The opposed mode positions the optoemitter and the optodetector opposite each other and aims the emitter directly at the detector. The retroflective sensing mode, also sometimes called the reflex mode, uses a retroreflective sensor containing both the photo-emitting and photo-detecting device in the same housing. The detectors sense reflected beams off the surface of, e.g., an object moving past which breaks these reflected beams.

The proximity mode uses a photosensor detecting the presence of an object directly in front of it. This sensor senses the photosensor's own reflected energy off the sensed object for proximity detection. This mode establishes a light beam, rather than detecting a broken beam. Some proximity sensors have a purposefully designed-in constraint limiting their sensing range. Sensing the difference between two received light levels makes contrast, or the light-to-dark ratio, very important. The following equation defines contrast:

$$\text{Contrast} = \text{excess gain (light condition)} / \text{(excess gain in darkness)}$$

Chapter 7

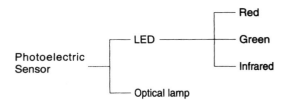

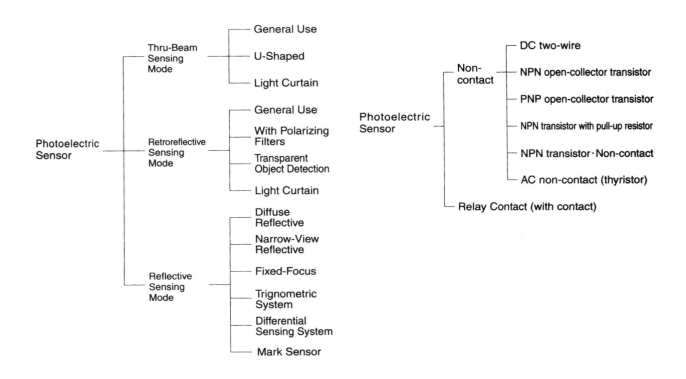

Table 7-1. An overview table grouping photoelectric sensors by various characteristics. (Courtesy of SUNX)

Defining Common Photoelectric Sensor Terms

A light curtain is light produced by a line or predefined array of emitting and receiving photoelectric sensors which detect objects anywhere in the defined sensing area.

Polarizing filters: theoretically, LED emitted light vibrates haphazardly; however, attaching a polarizing filter to the emitter and receiver allows the retroreflective sensor's emitter to send a horizontally polarized wave. Simultaneously, the receiver accepts only a vertically polarized wave. This arrangement allows sensing of solid objects and is

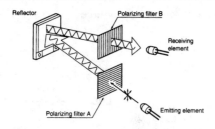

Figure 7-1. A two filter arrangement with a photoelectric sensor pair.

not affected by shape, color or surface material. Solid objects do not change the wave direction and returning light has the same wave direction as emitted light. (See *Figure 7-1*) The B filter in this figure purposely prevents stray light from the emitter from affecting the sensor's receiving element. Note the vertical and horizontal filter lines.

Transparent object detection is self explanatory, but is realized by using a reflector. A narrow view

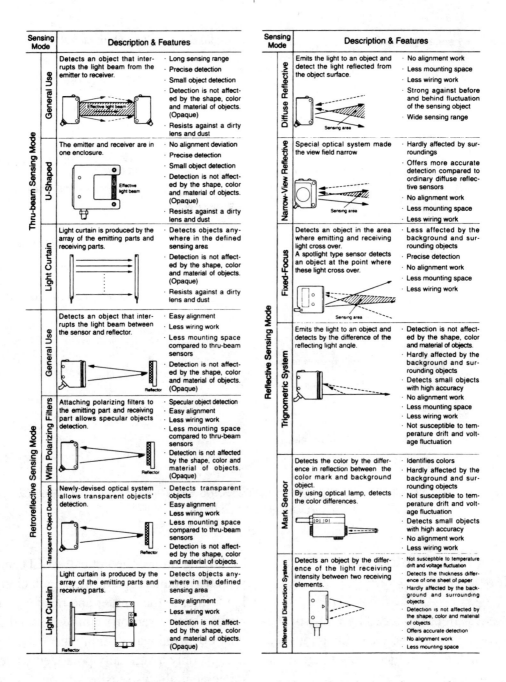

Figure 7-2. An illustrated summary of photoelectric sensing modes.

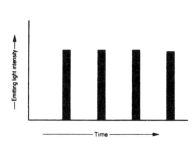

Figure 7-3. Modulated light. (Courtesy of SUNX)

flective photoelectric sensor has a narrow sensing area and is a more precise detection technique than ordinary diffuse reflective sensors. A fixed-focus sensor detects an object in the area where emitting light and receiving light cross each other. A spotlight sensor detects an object at the point where these lines cross. Background and surrounding objects do not affect this technique very much.

A trigonometric reflective sensor scheme emits light to an object and detects it by sensing the difference of the reflecting angle. This technique detects small objects with great accuracy and is impervious to shape, color, or sensing surface. It is also not susceptible to temperature drift or voltage fluctuations. A trigonometric reflective photoelectric sensor detects an object at a certain distance by returning light at an incident angle, independent of its reflection rate. This type sensor has a

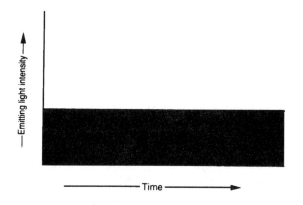

Figure 7-4. Unmodulated light.

Figure 7-5. The LA-510 and LA-511 IR laser emitter and controller. (Courtesy of SUNX)

light receiving lens through which returning light is generated according to the position where the returning light strikes the photodiode. *Figure 7-2* is an illustrated summary of optoelectric sensing modes.

Mark Sensors use an optical lamp to detect color by distinguishing between the difference in reflection between the object sensed and its background color. A differential distinction sensor or sensing system has two receiving elements (sensors) inside. It detects an object by measuring the difference of the light intensities experienced by the two receiving sensors.

Light Sensor Technologies

Photocells are most appropriate for good sensitivity to visible light, such as in ambient light or color registration. Photodiodes are best for tasks requiring a very fast response time and good light intensity linearly over several orders of magnitude.

Modulation Types

Most photoelectric sensors use pulse modulation (turning on and off) methods for emitting light. You can use this method on a light emitting source with very high switching speeds. Pulse modulation also allows the receiver to distinguish between the LED produced light and ambient light, while providing a long sensing range. (See *Figure 7-3*)

Class	Applicable Model No.	Degree of danger
Class 1	**LA-510** **LA-511**	Intrinsically safe for the operation
Class 2	—	Visible lights and weak output (wavelength 400 to 700 nm). Eyes are protected by their instinctive reaction to avoid the light when directly posed.
Class 3A	—	Dangerous when exposed to direct laser beam through optical devices. Output is less than 5mW for visible lights and max.5 times more than that of class 1 for other wavelength than visible lights.
Class 3B	—	Dangerous when exposed to direct laser beam. Observation of the pulse modulated laser radiation which does not have a focal point by diffuse reflection is not dangerous. The output that allows the safety laser observation is less than 0.5W.
Class 4	—	High power radiation Dangerous laser diffuse reflection may occur. It may leave harmful effects on the skin and become a cause of fire.

Table 7-2. Lasers classified by their intensities. (Courtesy of SUNX)

You can use unmodulated light emission in color mark sensors and high speed fiber optic sensors. This type of light has constant intensity (is unmodulated) and has high speed response times despite its slight susceptibility to ambient light induced disturbances. (See *Figure 7-4*)

Laser Photoelectric Sensors

There are four types of lasers:

1. Liquid pigment.
2. He-Ne, Ar and CO_2 gas lasers.
3. YAG, ruby and glass solid lasers.
4. Semiconductor lasers made from GaAs.

The lasers main advantages are that they have just one wavelength, are superior in directivity, have high energy density and easily interface with each other.

The type of photoelectric laser sensors which are most common for industrial and manufacturing applications are semiconductor lasers. These are laser photoelectric sensors which conform to class 1 in IEC Publication 825 and JIS C 6802, two industry standards. This makes stringent safety precautions unnecessary. *Table 7-2* shows a listing of four classes of lasers segregated by their intensities. *Table 7-3* lists lasers by type and wavelengths. *Figure 7-5* shows the SUNX LA-510 and LA-511

Chapter 7

TYPE OF LASER

Media creates Laser beam are as follows ;
Liquid ··················· pigment
Gas ···················· He-Ne, Ar and CO_2
Solid ··················· YAG, ruby and glass
Semiconductor ········ GaAs

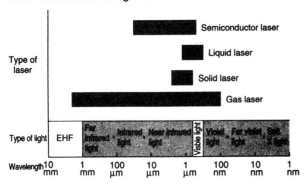

Table 7-3. Lasers classified by their type and wavelength zone. (Courtesy of SUNX)

IR laser emitter and detector with its LA-C1 controller. *Figure 7-6* is a laser's theory of operation. *Figures 7-7* and *7-8* show the IDEC MX1C miniature laser sensor and it feeding a PLC (Programmable Logic Controller) respectively. A PLC replaces old fashioned relays and electromechanical logic devices with semiconductor logic you can program or reconfigure in minutes.

Figure 7-9 shows an old ladder logic diagram with current flowing downward and only one switch and one solenoid activated. *Figure 7-10* shows a PLC's more modern equivalent circuit of *Figure 7-9*. Naturally, light sensors controlling these activations make a smarter, more flexible and reliable system.

A monitor is available which aligns the emitter and receiver to the best position, making light alignment easy, despite the laser's invisible nature. (See

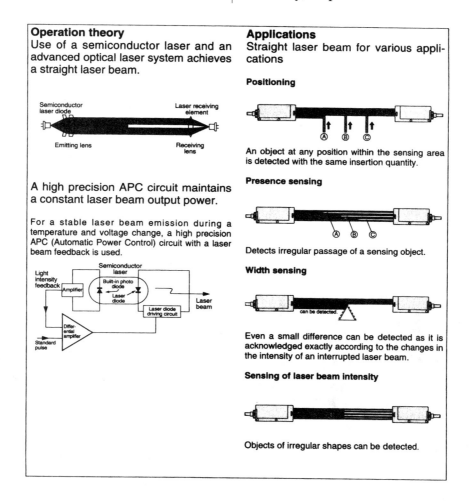

Figure 7-6. A laser sensor's theory of operation. (Courtesy of SUNX)

133

Figure 7-7. The IDEC MX1C miniature laser sensor. (Courtesy of IDEC)

Figure 7-11) A less expensive means to accomplish this is the infrared sensor card. See Appendix B for the source of supply. This card allows you to control the laser's shape since the projected laser causes a visible round spot on the sensor card sensors. The laser sensor consists of an emitter and a receiver (See *Figure 7-12*), a typical hookup of a SUNX model LA 510 photoelectric sensing laser pair. The accompanying laser controller ensures the presence of your application's amplitude signal. (See *Figure 7-13*) After accomplishing proper

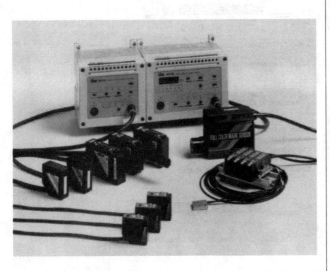

Figure 7-8. An IDEC PLC controller. (Courtesy of IDEC)

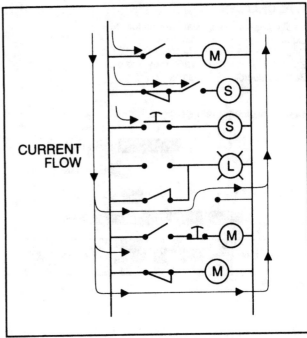

Figure 7-9. An old fashion electromechanical based ladder diagram. (Courtesy of SUNX)

alignment of the laser emitter and receiver you need to adjust the span. This means you must adjust the analog voltage to +5 volts while the receiver is at its full laser beam receiving potential. The

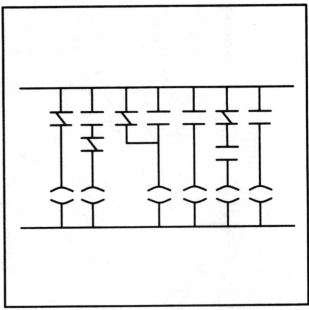

Figure 7-10. A modern PLC equivalent circuit of Figure 7-9. (Courtesy of SUNX)

Chapter 7

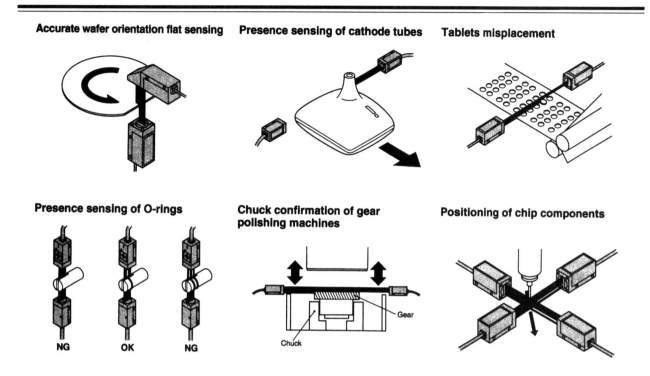

Figure 7-11. Specific applications of laser sensors, the NG in the figure is "No Good" and refers to either a missing O-ring or two O-rings when there should be only one. (Courtesy of SUNX)

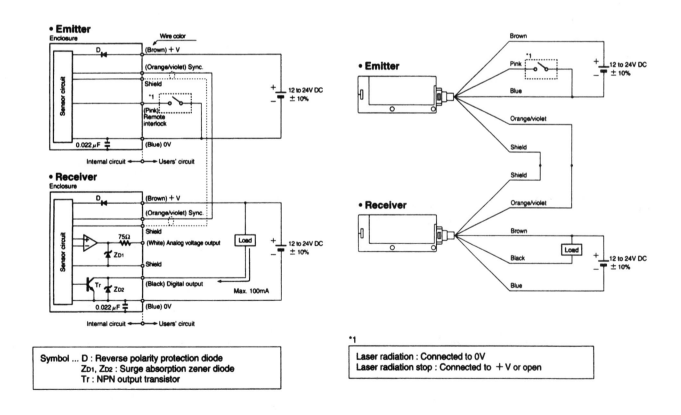

Figure 7-12. A typical hook up of a laser sensor. (Courtesy of SUNX)

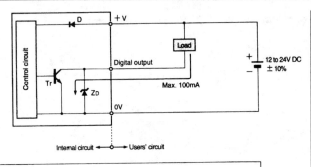

Symbol ... D : Reverse polarity protection diode
Zd : Surge absorption zener diode
Tr : NPN output transistor

Figure 7-13. *A laser sensor controller ensuring an amplitude signal's presence. (Courtesy of SUNX)*

old adjustment (See *Figure 7-14*) sets the digital output level.

Laser Sensor Precautions

Always ground the sensors' frame when using an electrically noisy switching power supply. A ca-

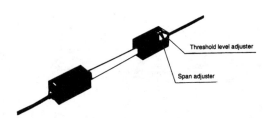

Threshold level adjustment
It sets the threshold level of digital output. When the threshold adjuster is turned clockwise, the threshold level of digital output increases.

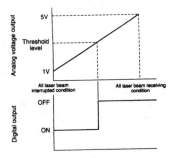

Figure 7-14. *Adjusting the digital output signal's threshold. (Courtesy of SUNX)*

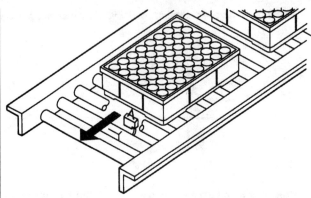

Figure 7-15. *An application exploiting the sensor's ON delay feature. (Courtesy of SUNX)*

pacitor grounds the sensors and improves noise resistance. If your application operates near an ultrasonic welder, its high frequency noises may interfere with the laser sensor's operation. This is despite the sensors' inductive metal mounting brackets. In these situations, ground the sensors to the mounting brackets. Also, do not use these laser sensors near steam or immerse them in water.

Self Diagnostics

These very sophisticated sensors can falsely diagnose themselves to show an improperly function-

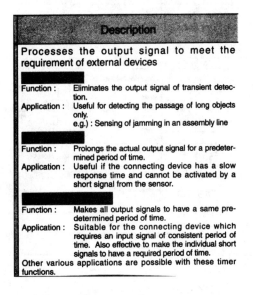

Table 7-4. *A listing of delay types. (Courtesy of SUNX)*

Chapter 7

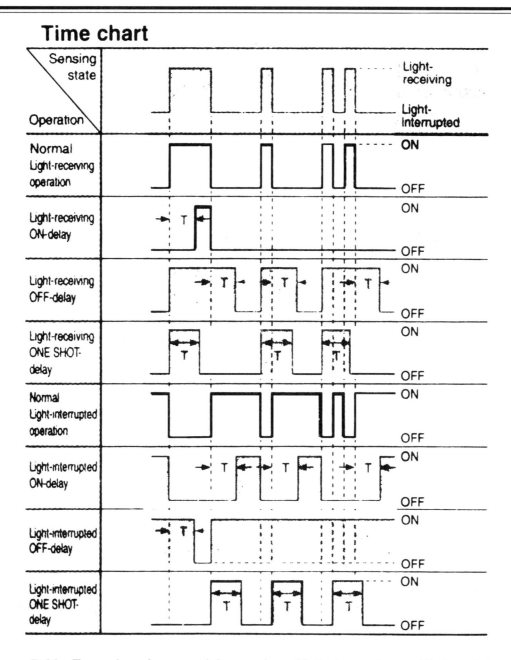

Figure 7-16. Examples of sensor delay modes with their sensing and timing states. (Courtesy of SUNX)

ing internal circuit. A dirty lens diminishes the light intensity the sensor's photodiode detector receives. Some sensors have an automatic sensitivity setting.

Types of Photoelectric Sensor Delays

These sophisticated photoelectric sensors have three primary types of delays. These are ON, OFF, and One-Shot delay. The ON delay purposely eliminates the output signal of transient detection. It is most useful in situations where you are detecting the passage of long objects or the jamming of a conveyer assembly line. (See *Figure 7-15*) *Table 7-4* lists types of delays. The OFF delay prolongs the actual output signal for a predetermined period. This is useful in applications in which the connecting device has a slow response time and you can't activate the sensor by a short pulse or short duration signal. The One-Shot delays make

137

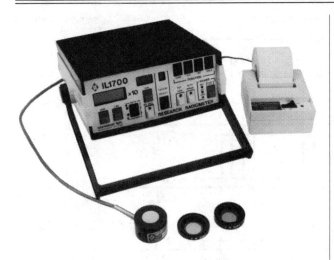

Figure 7-17. A radiometer with various wavelength sensors (probes). (Courtesy of International Light)

Fixed field proximity sensors do not response to an object's presence beyond its sensing range, no matter how reflective its surface is. These photoelectric proximity sensors use a lens or lens assembly to purposefully limit their sensing range and compare the amount of reflected light seen by two different optodetectors. The angle of arrival of reflected light is very critical with this type photoelectric sensor. There are trade-offs you'll have to make when considering a proximity photosensor. If you use such a diffused sensor with a lens or set of lenses you will experience a dramatic sensitivity to reflective surfaces and the angle of the light's arrival is very critical.

To reduce the proximity photosensor's sensitivity to specular (shiny or reflective) surfaces, you can use a proximity sensor with no lenses. These eliminate collimated sensing patterns due to the lenses, but the advantage is it shortens the sensing pattern and allows a much greater tolerance on the angle of arrival. However, if you need to sense a small

all sensor outputs have a predetermined period. This makes a sensor in this mode suitable for connecting to a device requiring an input signal of a consistent period. (See *Figure 7-16*)

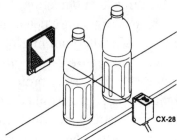

Figure 7-18. Applications of a sensor with built-in fluorescent light resistance. (Courtesy of SUNX)

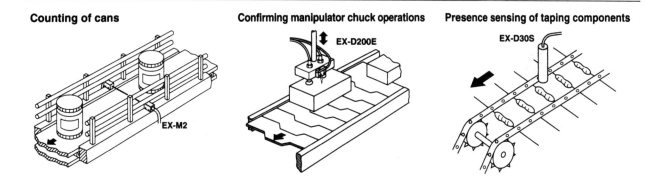

Figure 7-19. The SUNX EX-10 sensor in manufacturing applications. (Courtesy of SUNX)

object, it is better to use a photoelectric proximity sensor with lenses to concentrate the light and sensing to a very specific spot.

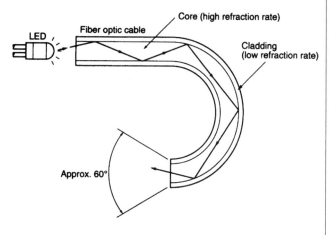

Figure 7-20. The fiber optic sensor's principle of operation. (Courtesy of SUNX)

Photometers measure visible incident light in footcandles or Lamberts/m^2. A **footcandle (fc)** is an older irradiance (flux/area) term. A Lambert/meter^2, a more modern term, equals 9.29 x 10^{-2} fc. A radiometer measures non-visible light, or IR and UV. These instruments have cosine corrected probes in front of the instrument (See *Figure 7-17*) which diminish light intensity as the cosine of the incident angle (or angle of arrival). The object to the right is its printer.

Applications of Unique Photoelectric Sensors

The CX20 optoelectric series of sensors from SUNX have a built-in inverter fluorescent light resistance circuit. This feature guards against malfunctions even under strong ambient fluorescent light which has several, usually three, different peak frequencies in its radiation spectrum. *Figures 7-18a* to *7-18d* are four applications using the CX20 series of sensors.

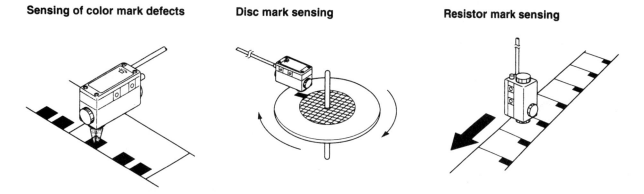

Figure 7-21. Optical lamp - fiber optics applications. (Courtesy of SUNX)

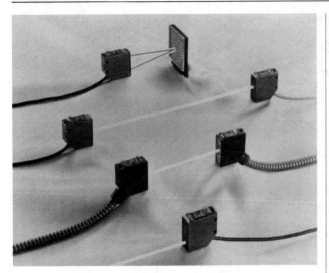

Figure 7-22. *The SUNX RX-LS200 trigonometric reflective sensor. (Courtesy of SUNX)*

The SUNX EX-10 series is the industry's smallest sensor, yet still has room for the built in amplifier. This small rugged photoelectric sensor withstands temporary water and nontoxic liquids submersion. One version can detect two different colors and has a fast response time of 0.5 milliseconds to accommodate a fast moving assembly line. *Figure 7-19* shows this sensor in three industrial manufacturing applications. The first counts cans. The next application senses if a manipulator chuck's operations conform to a predefined standard. The last application is a presence detection of taped electronic components on a reel.

Figure 7-23. *The SUNX model FX-7 fiber optic sensor with EEPROM memory. (Courtesy of SUNX)*

The LX series of optical lamp — fiber optics sensors from SUNX have an acute ability to differentiate between colors and a fast response time of 0.1 millisecond. *Figure 7-20* is the optical lamp/fiber optics principle of operation. It can detect the difference in, e.g., yellow and white wire and sense wire as small as 0.1 mm in diameter. Its optical adjuster, once properly set, makes this sensor virtually impervious to variations in supply voltage and ambient operating temperatures. If a motor has a marked rotating shaft, this sensor can detect and count the revolutions in a specific interval, such as RPM. *Figure 7-21* shows three electronics applications in which this photoelectric sensor either detects a mark and/or a color mark. This series of photoelectric sensors has several characteristics making the applications in *Figure 7-21* possible. The sensor's high sensitivity to color detection closely resembles that of the human eye (they are CIE curve corrected). Their high speed response is 10 μseconds. This allows a maximum frequency of 100 kHz. You can also change the lens' mounting position. This allows sensing from both the top and the side.

Real World Photoelectric Sensors

The SUNX RX-LS200 is a trigonometric reflective sensor. (See *Figure 7-22*) The light's receiving angle, not intensity, determines this sensor's range. In its fixed focus mode, the sensor detects an object in the overlapping area of emitting and receiving light. This reflective ratio determines sensing performance. In its diffuse sensing mode, the sensing range differs by this ratio and by an object's size.

Figure 7-23, the SUNX model FX-7 fiber optic photoelectric sensor, has nonvolatile automatic sensitivity settings. Its internal EEPROM provides an OFF delay timer. An ON/OFF button allows sensing and non-sensing. A switch for changing frequencies prevents crosstalk. This is handy when mounting these sensors next to closely spaced fiber cables. *Figure 7-24* shows three electronics manufacturing applications of this sensor.

Chapter 7

Wafer presence sensing
Detects wafers without being affected by color, glossiness or transparency of the sensing object surface when combinating with fixed-focus reflective fiber optic cable.

Wafer sensing in vacuum chambers
Detects wafers in vacuum chambers.

Front and back judgment of chip components
Checks front and back of small chips such as a 1005 chip by using a pinpoint spot lens.

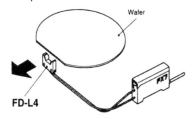
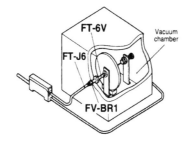
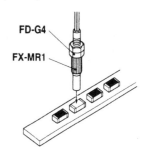

Figure 7-24. Three electronics manufacturing applications of the SUNX model FX-7 fiber optic sensor. *(Courtesy of SUNX)*

Chapter 7

Chapter 7 Quiz

1. A proximity sensor:
   A. Senses an object directly in front of it.
   B. Senses the photosensor's own reflected energy.
   C. Establishes rather than breaks a light beam.
   D. All the above.
2. A light curtain is a line or _____ which detect an object anywhere in its defined sensing area.
   A. Random array of emitting and receiving photoelectric sensors.
   B. Predefined array of emitting and receiving photoelectric sensors.
   C. Both A and B.
   D. Neither A or B.
3. Which type photoelectric sensor detects where emitting and receiving light intercept?
   A. A fixed-focus sensor.
   B. A trigonometric sensor.
   C. A mark sensor.
   D. None of the above.
4. Which type photoelectric sensor emits light to an object and senses the difference angle of the received light?
   A. A mark sensor.
   B. A sensor with a polarizing filter.
   C. A trigonometric reflective sensor.
   D. None of the above.
5. Mark sensors:
   A. Typically sense colors.
   B. Sense color by distinguishing between the reflections of an object sensed and its background color.
   C. Both A and B.
   D. None of the above.
6. Most photoelectric emitting sensors use what?
   A. Pulse-modulation.
   B. No modulation.
   C. Phase-modulation.
   D. AM modulation.
7. A photodiode has:
   A. Faster response times than a photocell.
   B. Better light intensity linearity over several orders of magnitude.
   C. Less sensitivity to visible light than a photocell.
   D. All the above.
8. A laser has:
   A. Just one wavelength.
   B. Excellent directivity.
   C. A high energy density.
   D. All the above.

9. You should never use laser sensors:
   A. Around steam.
   B. In the open air.
   C. In high temperature environments.
   D. All the above.
10. How do you reduce a proximity sensor's sensitivity to specular or shiny objects?
    A. Use a two lens double layer.
    B. Use it with no lens.
    C. Lengthen its sensing range.
    D. Use a color tinted lens.
11. A photometer measures irradiance of visible light, T or F?
12. A radiometer measures irradiance of visible light, T or F?
13. Some photoelectric sensors have a built-in inverter which guards against malfunctions in fluorescent and ambient light, T or F?
14. Optical lamp — fiber optics sensors have a poor ability to tell the difference between colors, T or F?
15. Some sensors have nonvolatile EEPROM memories which allow changing frequencies which helps produce crosstalk, T or F?

Chapter 8
Experimenting With Modern Optoelectronic ICs

Chapter 8
Experimenting With Modern Optoelectronic ICs

This chapter covers four representative modern optoelectronic ICs from T.I. (Texas Instruments) and shows how you typically use them. These are:

1. The TSL250 and TSL260 series of light-to-voltage optical sensors.
2. The TSL214 64 x 1 integrated opto sensor array.
3. The TSL230 programmable light-to-frequency converter.
4. The T.I. TMC3637 remote control transmitter and receiver IC.

This is not an IR LED optoelectronics IC, but attaching an IR LED to its open drain output configures it as both an unmodulated and an IR modulated transmitter. Attaching a phototransistor on its input makes it an IR receiver.

A Photodiode's Role in Light-to-Electrical Conversion

The most fundamental part of optical signal conditioning is amplifying incident ambient light and converting it into electrical digital signals. A silicon photodiode most easily measures real time

TSL250/TSL260 Light to Voltage Sensor

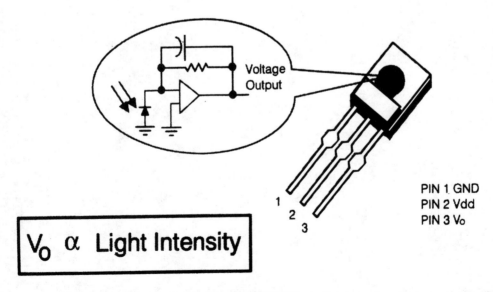

Figure 8-1. The TSL260 black plastic case's IR filter, pin out, physical case profile, and light-to-electrical signal transformation op amp. (Courtesy of Texas Instruments)

Chapter 8

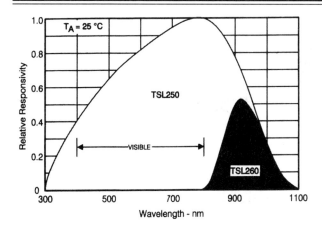

Figure 8-2. The TSL250/260's spectral sensitivities. (Courtesy of Texas Instruments)

light intensity. A photodiode's active surface receives incident radiation and its silicon absorbs it. This generates hole-electron pairs. These produce a photo current across a reverse-biased P-N junction. Photodiodes serve as the intentionally exposed light gathering base for phototransistors. Phototransistors have no base pin like an ordinary transistor, just an exposed light sensitive gathering surface for their base. With an appropriate op amp (See *Figure 8-1*), the photodiodes' measure light levels are converted into electrical digital or analog signals. Note the paralleled RC feedback components in *Figure 8-1*.

The TSL250/260 Series ICs

These are the least complex ICs we examine is the TSL250/260 series of light-to-voltage sensors.

These measure low light levels in electrically noisy environments. They have a large area photodiode and a transimpedance amplifier for converting light induced photodiode current to voltage.

T.I.'s basic semiconductor process allows photodiode spectral sensitivity to span a wide 400 to 1,100 *nm* range. Their cases' plastic encapsulation though acts as an optical filter, purposely narrowing this spectral sensitivity range. Specifically, the TSL260 series has an IR only transmissive coating which filters out all light except IR. (See *Figure 8-1* again) The main difference in this series of ICs is the photodiode's light sensing area and the op amp's feedback resistors.

The TSL250 series of T.I. ICs are visible light-to-voltage sensors. The TSL260 series are virtually identical ICs, except they operate in the IR range. (See *Figure 8-2*) We'll use the TSL260 as our example. See *Figure 8-1* again for pin out and the physical outline package. The bulging bubble on the IC's surface is naturally the light sensor area and its black plastic case serves as an IR filter. This IC's applications greatly vary, but placing a pull-up resistor on its output (See *Figure 8-3*) enhances performance by extending linear operation to near V_{DD} with minimal effect (several millivolts) due to dark voltage, the voltage with no light input. *Figure 8-4* is a simple application in which the output goes high when you interrupt the light beam. This operates at a distance of several inches and serves as an optical interrupter switch or reflective

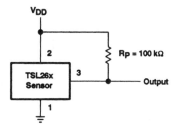

NOTE: Pull-up resistor extends linear output range to near V_{DD} with minimal (several millivolts typical) effect on V_{DARK}; particularly useful at low V_{DD} (3 V to 5 V).

Figure 8-3. Extending the TSL260's linear operation. (Courtesy of Texas Instruments)

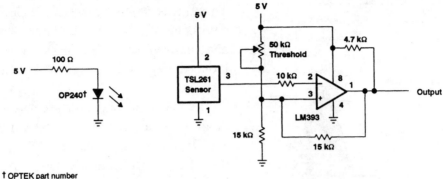

† OPTEK part number
NOTE: Output goes high when beam is interrupted; working distance is several inches or less. Intended for use as optical interrupter switch

Figure 8-4. A TSL260 optointerrupter short range application. (Courtesy of Texas Instruments)

object sensor. *Figure 8-5* is a similar application of a pulsed optical beam interrupter in which the output pulses are low, until you interrupt the light beam. This has a far greater useful range, up to 20 feet when you use lenses. This circuit is ideal for door openers. *Figure 8-6* is a proximity detector and its output goes low when light pulses reflect back to the sensor. The range is 6 to 18 inches, which depends on the object's reflectance factor.

If you use this IC as an IR transmitter and receiver pair, you activate the transmitter in the top illustration in *Figure 8-7* by pushing the momentary contact push-button switch, S1. This enables the 555 timer which pulses IR diode D1. The TSL260 IR sensor senses this and feeds an NE567 tone decoder. Its output goes to a flip-flop with its Q output inverted by the current sinking transistor, Q3.

This transistor conducts (is active with) a logic low on its base, which draws current through its collector to the emitter. This current also flows through U3, an optocoupler, which turns on the optical triac driver, Q2. This activates, or remotely controls, an AC device, such as a ceiling fan or lamp plugged into its power terminals, J1.

The TSL214 Series IC

The TSL214 is a 64-element line sensor with each pixel on 125 mm centers. These are charge sensors or devices which develop charges on pixels (addressable dots) proportional to the product of light intensity times exposure time. This is analogous to photographic film experiencing light exposure. This IC only requires a +5 volt supply and pixel output clock pulses. It typically operates in

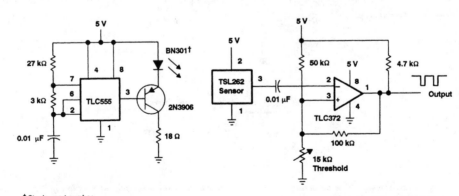

† Stanley part number
NOTE: Output pulses low until beam is interrupted. Useful range is 1 ft to 20 ft; can be extended with lenses. This configuration is suited for object detection, safety guards, security systems, and automatic doors.

Figure 8-5. A greater range TSL260 optointerrupter application. (Courtesy of Texas Instruments)

Chapter 8

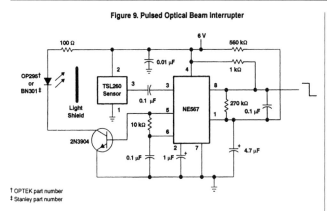

† OPTEK part number
‡ Stanley part number

Figure 8-6. A TSL260 proximity detector.
(Courtesy of Texas Instruments)

the 10 kHz to 500 kHz range. This design is an alternative to either discrete photosensor arrays or CCD (Charge Coupled Device) line imagers. A discrete photodiode's active surface area is about 1,000 mm and a CCD has about a 10 mm active surface area. At 125 *mm*, this IC is a practical happy medium in resolution, cost, and complexity.

Figure 8-8 is the IC's block diagram. Typically the T.I. TSL214 and the 128 pixel TSL215 operate in concert with a light source, such as a DC powered constant current LED or an incandescent lamp and a microcontroller. The microcontroller allows such

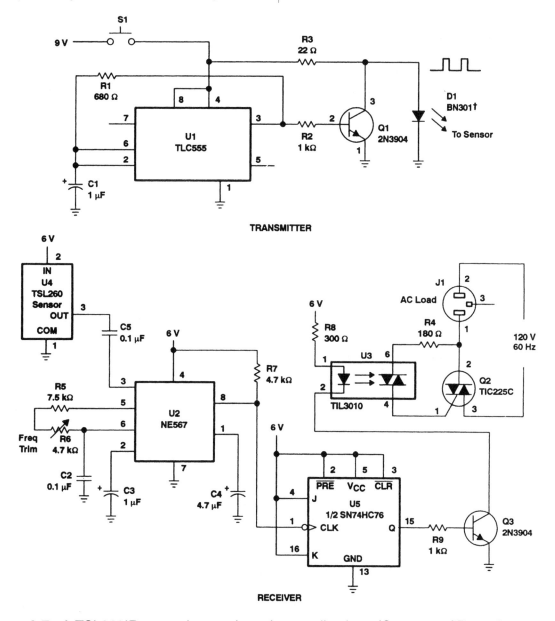

Figure 8-7. A TSL260IR transmitter and receiver application. (Courtesy of Texas Instruments)

149

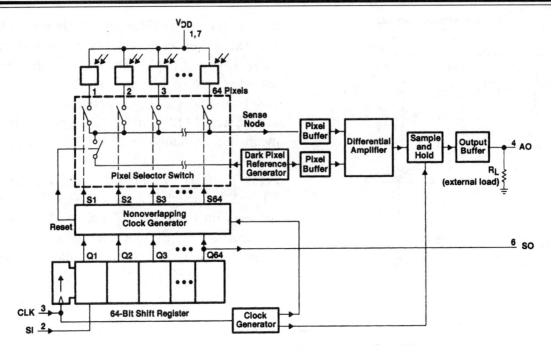

Figure 8-8. The TSL214 block diagram. (Courtesy of Texas Instruments)

features as auto-calibration. The TSL213, configured in such a system, can detect objects' edges and positions, light-to-dark transitions, and pixels. The pixel level detection function converts the level of each pixel to a binary state on a processor's output port. This creates a digital logic compatible signal. The conversion occurs in real time and is a digitized version of the sensor's output, taking cognizance of pixel intensity.

The TSL230

This light-to-frequency IC does not use a traditional A/D conversion technique. It uses a switched capacitor current-to-frequency converter with an output you detect with either a frequency counter or a period timer. (See *Figure 8-9*) The IC has a symmetrically configured (10 x 10) 100 photodiode matrix array. You can selectively activate all 100, 10, or just 1 photodiode. This accommodates a wide range (8 orders of magnitude) of incident light from 0.001 to 100,000 mW/cm^2. Otherwise, using all 100 photodiodes under high ambient lighting conditions quickly saturates the TSL230 IC. This causes premature and inaccurate attainment of its 1 MHz limit. This 8-pin mini-DIP IC has two additional control lines with which you scale its output frequency for a fixed pulse width divided by 2, 10 or 100. This could prove helpful in interfacing with pulse window circuits. Chapter 9 discusses a light intensity meter built with this IC.

Chapter 8

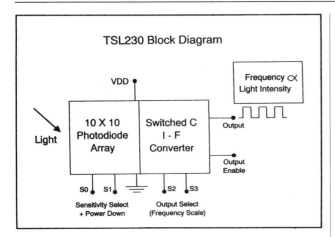

Figure 8-9. The TSL230 block diagram. (Courtesy of Texas Instruments)

A Typical Light Modulation IC

Figure 8-10 is a block diagram of the Hamamatsu S3599, an older but typical light modulation IC, with its truth table. You externally connect an IR LED to this IC to configure synchronously detecting photoreflectors and optointerrupters. The S3599 allows background light of 10,000 lux or 5,000 lux minimum at 2856 K with a minimum detection level of 1 μW/mm² at λ = 940 nm. This detection level is 1,000 times less than the TSL230, a more modern IC. This S3599 is admittedly not as compact as the T.I. series ICs but provides greater flexibility by allowing selection of external components. This is a practical example of making a trade-off in parts count to an application's flexibility.

The following five key elements typically comprise such an IC:

1. The oscillator/timing signal generator circuit's reference oscillator output results from charging and discharging the built-in capacitor with a constant current. The oscillator output feeds the timing signal, creating an LED drive pulse, plus various timing pulses for digital signal processing.
2. The LED driver circuit directly drives the LED with constant current pulses (power strobes it), using the LED drive pulses created by the timing signal generator circuit. The duty cycle is 1/16.
3. The photodiode/preamplifier circuit uses the photodiode formed on the same chip with a peak spectral sensitivity of 800 nm. The photocurrent of the photodiode is converted to a voltage through the preamplifier circuit. An amplifier circuit in the preamplifier expands the dynamic range in response to DC or low-frequency background light. It also boosts the signal's detection sensitivity.
4. The capacitive coupling buffer amplifier reference voltage generator circuit eliminates low-frequency background and DC offset within the preamplifier section by capacitive coupling. The buffer amplifier amplifies up to the comparator level which the reference voltage generator produces.
5. The comparator circuit has a hysteresis function, preventing chattering from small fluctuations in input light.

The TMC3637

This compact 8-pin IC is the most complex IC reviewed. It has numerous modes, including a test mode and registers for its addressable internal

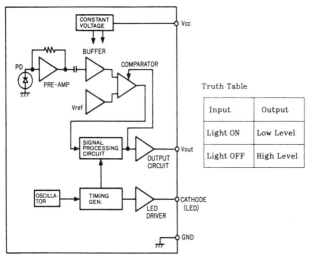

Figure 8-10. The Hamamatsu S3599 block diagram.

Functional Block Diagram

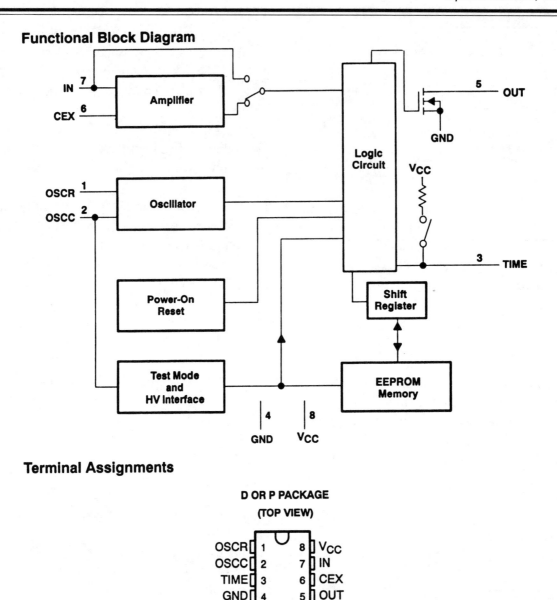

Figure 8-11. The TMC3637 block diagram. (Courtesy of Texas Instruments)

EEPROM; however, it does not require any external DIP switches. (See *Figure 8-11*) The IC continuously monitors and decodes its internally stored security code at over a 90 kHz rate. You can only alter these over four million codes with a TMC3637 programming station, see this IC's data manual. This IC uses 2 μm semiconductor processes for low power consumption and has an internal oscillator, alleviating the need for an external time base or crystal. It also has 48 possible configurations as a transmitter and 14 possible configurations as a receiver. You can transmit in a single, multiple, or continuous manner. The TMC3637 data manual explains all this in detail. The most difficult aspect of this IC is programming its wide EEPROM.

There are both EEPROM read and write modes and there are nine of the 31 EEPROM cells, CA through CI, which determine the transmitter's or receiver's multitude of configurations. There are

Chapter 8

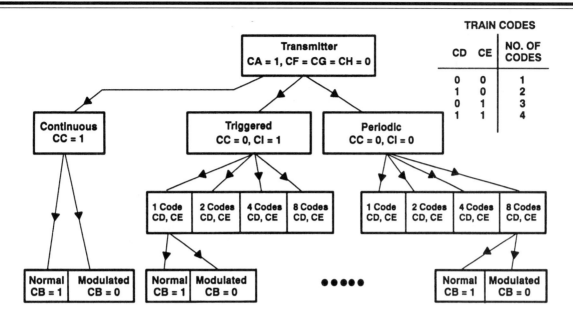

Figure 8-12. A TMC3637 IC's possible transmitter configurations. (Courtesy of Texas Instruments)

three transmission modes (not to be confused with configurations):

1. The normal continuous mode.
2. The triggered mode.
3. The periodic transmission mode.

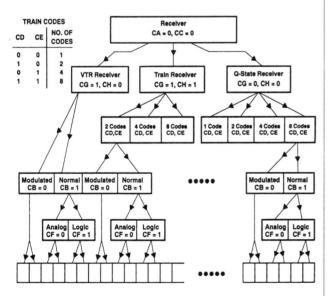

Figure 8-13. A TMC3637 IC's possible receiver configurations. (Courtesy of Texas Instruments)

An IR-Based TMC3637 Transmitter

The TMC3617 uses its internal clock of approximately 26 kHz to clock out the data. There are 31 data bits which the EEPROM stores and the last of these bits determine one of 44 possible transmission or one of 14 possible receiver configurations. (See *Figures 8-12* and *8-13*)

An IR-Based TMC3637 Receiver

The IR phototransistor based receiver (See *Figure 8-14*) is in the inverting mode. Using OUT on the transmitter to send code is considered inverted. This receiver is in the logical normal (1-code) Q-state inverting receiver configuration with EEPROM cells CA to CI programmed as: 0 1 0 0 0 1 0 0 1. The signal path between the transmitter and receiver does not invert the signal. Using the logic mode CF = 1 also does not invert the signal. The result is inverted at the controller of the receiver; therefore, we use CI = 1 for an inverting receiver. The oscillating frequency of the receiver is approximately ten times as great as that of the transmitter or 260 kHz. OUT on the receiver maintains the same status for about 0.5 sec. or 1 MΩ x

153

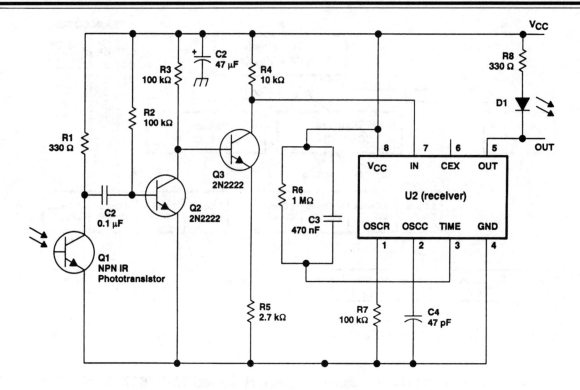

Figure 8-14. A TMC3637 IR based receiver. (Courtesy of Texas Instruments)

470 nF. See the receiver's R6 and C3 RC tank or parallel combination from the TIME pin 3 to V_{CC}, pin 8.

IR Coupled Modulated Transmission Receiver

Figure 8-15 is an IR receiver working in the modulated continuous configuration. You may also use this modulated receiver with a normal IR transmitter provided the EEPROM's CA through CI cells are: 1 0 1 0 0 0 0 0 0. The transmitter's oscillating frequency must always be 120 kHz and you do this by selecting the correct combination of R_{osc} and C_{osc}. The $f_{osc} = 5/(4 \times R_{osc} \times R_{osc})$ where C_{osc} is the capacitor from OSCC to GND and R_{osc} is the resistance value from OSCR to GND. Cascade the TMC3637 with a TDA3048 demodulator IC or its equivalent to process and demodulate the signal. The receiver is configured as a modulated (1-code) Q-state inverting receiver with its EEPROM CA to CI cells as follows: 0 0 0 0 0 0 0 0 1. The oscillator frequency of the receiver is about 900 kHz.

Chapter 8

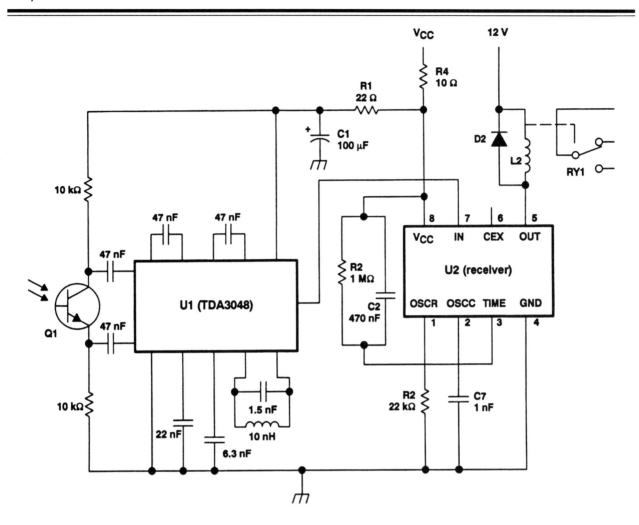

Figure 8-15. A TMC3637 IR coupled receiver. (Courtesy of Texas Instruments)

Chapter 8 Quiz

1. When light strikes a photodiode's silicon surface, the silicon rejects the light, T or F?
2. Photodiodes have a base pin, just like ordinary transistors, T or F?
3. Texas Instruments uses a 2 mm process to achieve lower power consumption, T or F?
4. Texas Instruments purposely narrows its optoelectronic IC's broad spectrum by, e.g., making the TSL260's black plastic an IR filter, T or F?
5. Using a pull-up resistor with the TSL260 enhances its operation by extending the linear operation almost down to V_{DD}, T or F?
6. *Figure 8-5* would make an ideal garage door opener, T or F?
7. The Hamamatsu S3599 can detect light with 1,000 times less intensity than the T.I. TSL230 can, T or F?
8. The TSL230's frequency output is proportional to light intensity, T or F?
9. The easiest aspect in working with the TMC3637 is programming its EEPROM, T or F?
10. The oscillator frequency of the TMC3637's IR based receiver is about 100 times that of the transmitter, T or F?

Chapter 9
Projects: Experimenting with the TSL Series of Optoelectronics IC

Chapter 9
Projects: Experimenting with the TSL Series of Optoelectronics IC

This chapter will experiment with the TSL series of intelligent optoelectronic ICs from Texas Instruments. Should you elect to purchase this evaluation board, it contains three of these sensors (optoelectronic ICs), and all necessary ancillary external hardware:

1. An IR filter cap for one of these three ICs, the TSL230.
2. All resident on board drive and evaluation software.
3. A software disc.
4. A cable.
5. Connectors such as a subminiature DB-25 25 pin-to-telephone jack conversion connector which allows you to use it with your PC's serial port.
6. Two pairs of matched wavelength light detectors and sources. One pair responds to visible light and the other matched pair responds to IR.

This kit supplies everything to build five projects, primarily based on the intelligent light-to-frequency converting opto-IC, these are:

1. The photometer.
2. A light meter.
3. The heart rate monitor.
4. The proximity detector.
5. The datalogger for data dumps via your PC's serial port.

If you do not wish to purchase this board, there is a stand-alone project in this chapter, and another in Chapter 10, both built around the TSL230 opto-electronics IC. This 8-pin mini-DIP IC has a 10 x 10 array of photodiode sensors built on it, which you can enable as:

1. A single photodiode.
2. 10 photodiodes.
3. 100 photodiodes.

You can also scale the frequency output like a programmable counter IC. The experiments allow you to make trade-offs in sensitivity, resolution, and the number of the 100 photodiode sensors you enable. Enabling too many photodiodes, in a strong ambient environment, saturates the internal light-to-frequency conversion circuits. The TSL230 IC comes with various part numbers, differing only in their suffixes. If no suffix exists, that indicates a 20% tolerance in output frequency, an A suffix indicates a 10% tolerance and a B suffix indicates a 5% tolerance.

If this IC intrigues you, you can order an evaluation kit, the TSL230EVM, from T.I. for $79.95, phone 214-638-0333. The IR blocking glass optical filter is from Hoya, the camera lens filter manufacturer. The filter is a cap which physically encases the entire TSL230 opto IC but has a small aperture in the top, allowing light to enter. It blocks IR to the TSL230 IC so it can more closely simulate the human eye's response. It also permits light detection down to $l = 300$ nm.

Chapter 9

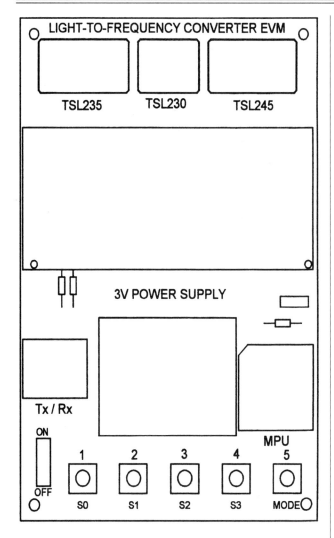

Figure 9-1. The TSL230EVM board's outline. (Courtesy of Texas Instruments)

This experimenter's board (See *Figure 9-1*) comes with Windows™ based software. The software provides you with a vivid presentation of the TSL230 connected to a counter, with toggle switch icons. These provide the input voltage levels to the programming lines. On screen displays help you know, as well as control, your operating mode. It also presents the realistic real-time on screen counter's frequency output.

In the photometer application, you can click between Lux and footcandles as the light output unit symbols, and also select between autorange and manual operation. The evaluation board has two photodetectors and wavelength matching sensors.

The visible light sensor pair, the TSL235, are approximately 0.5" apart. Placing your finger between them determines your pulse rate by detecting changes in your finger's blood flow. (See *Figure 9-2*) The TSL245 is an infra-red (IR) light photodetector and source LED pair. *Figure 9-3* shows the block diagrams of the two wavelength matched LED pairs. They are still light-to-frequency converters; however, they lack the TSL230's sophisticated controls. (See *Figure 9-4* for a comparison.) This shows:

1. The TSL230's block diagram.
2. The two groups of selectable options provided by control pins S1/S2 and S3/S4.

It also shows the IC's eight pins and their functions. Pins S1 and S2, the sensitivity select pins, allow you to enable just one, 10, or all 100 photodiodes. *Figure 9-5* shows the effects of the TSL230's programming pins, S0 and S1.

To quickly summarize the capabilities of both the TSL235 and TSL245, let's examine *Figures 9-6* and *9-7*. These show:

1. The TSL245 IR LED pair's spectral response, which is radiant, non-visible light.

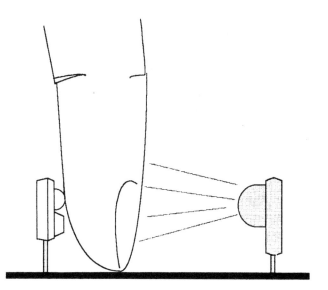

Figure 9-2. Measuring your heart rate through your finger. (Courtesy of Texas Instruments)

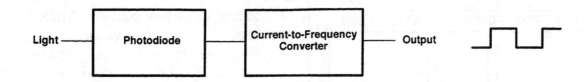

Figure 9-3. *The TSL235 and TSL245's simple block diagrams. (Courtesy of Texas Instruments)*

2. The TSL235's output frequency versus its irradiance. Despite their ostensible simplicity, you can interface both these sensors to a microprocessor. (See *Figure 9-8*)

For maximum accuracy, use period measurements and a fast reference clock with resolution directly proportional to the reference clock's rate.

You may want to enable only 1 or 10 photodiodes in the presence of a high intensity light source. Otherwise, this bright light might overwhelm or spot saturate the photodiodes, driving them into saturation. The following equation quantitatively describes this excess gain as simply:

Excess gain = (light energy on receiving element/sensor amplifier's threshold)

This IC's ability to select 1, 10, or all 100 photodiodes provides, in effect, an electronic iris or the equivalent of a camera lens's aperture control. The S3 and S4 pins provide four combinations of frequency output. These yield either:

1. The full undivided output frequency.
2. A divide-by-2.
3. A divide-by-10.
4. A divide-by-100.

The OE pin places the IC in a high-impedance or tri-state mode, disconnecting it from a bus operated system.

To obtain optimum performance, maintain a 5 or a 3.5 volt supply on the TSL230. At 4.5 volts, there is a 1.015 normalized output frequency, which

Terminal Functions

TERMINAL NAME	NO.	I/O	DESCRIPTION
GND	4		Ground
OE	3	I	Enable for f_O (active low)
OUT	6	O	Scaled-frequency (f_O) output
S0, S1	1, 2	I	Sensitivity-select inputs
S2, S3	7, 8	I	f_O scaling-select inputs
V_{DD}	5		Supply voltage

Selectable Options

S1	S0	SENSITIVITY
L	L	Power Down
L	H	1×
H	L	10×
H	H	100×

S3	S2	f_O SCALING (divide-by)
L	L	1
L	H	2
H	L	10
H	H	100

functional block diagram

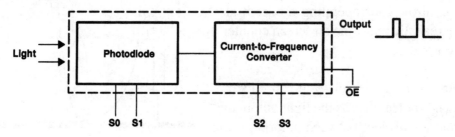

Figure 9-4. *The TSL230's block diagram, control lines, pin functions and truth tables for both sensitivity and output frequency scaling. (Courtesy of Texas Instruments)*

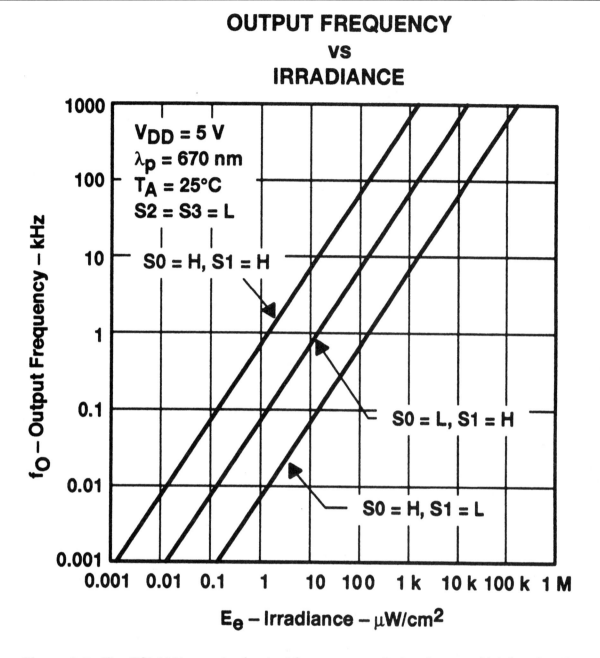

Figure 9-5. The TSL230's graph of output frequency vs. its irradiance, which is a function of pin S0 and S1 states. (Courtesy of Texas Instruments)

equates to a 1.5% error. But keep in mind this IC already has an output frequency tolerance which is either 20, 10, or 5% depending on the part number's suffix of either none, A, or B respectively.

Precautions

Fluorescent lights are the most common source of AC light noise (flicker) in the transmissive (Mode 2) or reflective (Mode 3) measurements. This can cause erratic readings when making ambient light measurements in Mode 1. Since this type of lighting is common in office and lab environments, it is difficult to avoid these problems when using the TSL230EVM evaluation board.

A common technique to minimize these effects is to sample the sensor at an interval which is an ex-

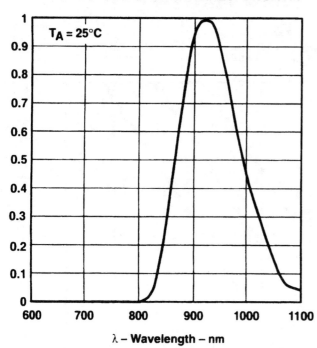

Figure 9-6. *The TSL245 IR sensor pair's spectral response. (Courtesy of Texas Instruments)*

act multiple of the period of the fluorescent light flicker (120 Hz or 8.33 ms). This takes the readings at the same point on the AC waveform each time. This simplifies the problem to a mere DC offset. This is exactly what you do in your light synchronizing experiment.

A small deviation in the sampling interval causes a beating effect in which flicker erroneously presents itself as a slowly changing ambient light reading. The first precaution is to shield, as much as possible, any transmissive or reflective experiments from ambient light. The alternative is to synchronize the sampling interval with the flicker frequency. This technique exists in the TSL230EVM software.

Sleep Mode

The TSL230EVM automatically enters the power-saving sleep mode if you don't press the buttons for three minutes. Press any button to wake up the TSL230EVM. Although power consumption is lower in the sleep mode, you should switch off the power to conserve the batteries when not using the TSL230EVM.

Mode 1 - TSL230 Evaluation

Mode 1 allows evaluation of the on-board TSL230 IC in general lighting measurement situations. Buttons 1 through 4 toggle the state of the programming lines S0 through S3, respectively. The display indicates the current programmed state of the TSL230. On power-up, the evaluation board defaults to 1x sensitivity and divide-by-100 output scaling. This corresponds to the states H, L, H and H for pins S0, S1, S2, and S2, with "H" a logic high and "L" a logic low. The nine left-most display resulting changes in the TSL230's configuration. The right-most five digits indicate the TSL230 output frequency in Hz.

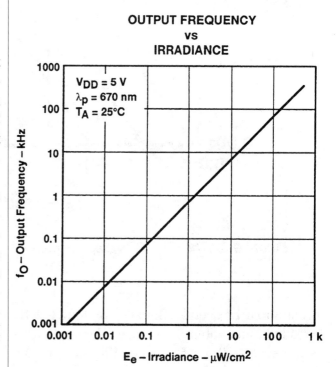

Figure 9-7. *The TSL235's graph of output frequency vs. its irradiance. (Courtesy of Texas Instruments)*

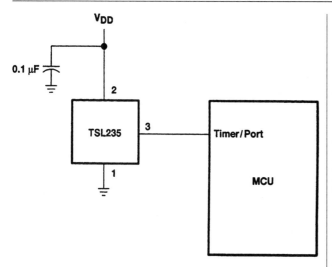

Figure 9-8. Interfacing a TSL235 or TSL245 to a microprocessor.
(Courtesy of Texas Instruments)

Hardware Design

Two system-level constraints dictate performance limitations of the system:

1. The board must operate on two 1.5 volt batteries.
2. A minimum component count.

An 8-bit microcontroller, with a 16-bit on-board timer and input-capture function, controls frequency measurement and system features. It uses period timing, instead of the frequency measurement technique, to exploit the processor's timer capture function. The timer hardware constrains the system's dynamic range. It is 65,535 (maximum number of counts) divided by 10 (approximate minimum number of counts due to software overhead latency in the timing loop) or a 6,553:1 ratio, (approximately 72 dB). This is the equivalent of a 12-bit A/D converter.

The 3 volt operation constraint places an upper limit of 2 MHz on the microcontroller clock. This translates into a 4 ms resolution (due to the internal divide-by-8) for period measurement, with a lower limit of 65,535 x 4 µs, or 262 µs (3.8 Hz) and an upper limit of 10 x 4 µs, or 40 µs (25 kHz).

Unlike typical designs, this data acquisition system uses no A/D converters. You obtain the best accuracy with the period measurement technique at the lower end of the input range (where the number of counts per period is high) rather than at the upper end (where the number of counts is low). For greatest measurement accuracy, you should keep signals out of the upper end of the input range. For example, for the TSL230EVM evaluation board to maintain a measurement accuracy of 1% or better, you must keep the signal below 2.5 kHz (100 x 4µs = 0.4 ms, 1/0.4 ms = 2.5 kHz).

Measurement time in this system is the period of the input signal plus an additional constant amount of time for software overhead and transmission of the data through the serial port. In order to maintain a constant sampling rate, you must also keep the signal's period well under the sampling period. This limits the system's low light measurement level. If you cannot meet this requirement, the signal may be recoverable only if you know the signal's frequency range and amplitude relative to the other frequency components.

Conventional D/A systems use low-pass filters on the input of A/D converters to eliminate frequency components above one-half the Nyquist frequency (the sampling rate). This eliminates the phenomenon known as aliasing, in which frequency components at greater than 1/2 the sampling rate appear as lower frequency components in the recovered data. In the case of the TSL-family of light-to-frequency converter ICs, it is not possible to use an anti-aliasing filter since the chip already contains the circuitry. You must observe the Nyquist criteria to accurately recover the waveform of a modulated light source. The sampling rate [1/(output period + overhead)] must be at least twice the frequency component in the light signal.

The design of a light measurement system using a light-to-frequency converter is straightforward. Most of the power and functionality lies in the software. For simplicity, some necessary or desirable functions, such as power supply regulation, voltage supervisors, and dedicated LED drivers do not exist on the EVM board. The low-dropout regulators with current-limiting serve the LCD module well since it requires 5 volts, and the battery is only 3 volts. You must obtain this higher voltage from the MAX218 RS-232 driver. Since we use only one of two transmitters, the 6.5 volts for RS-232 transmission remains available at low-current from the V+ line on the MAX218. A conventional low-dropout regulator to generate the LCD's 5 volts suffers on power-up from the PNP pass transistor and generates a transient high-current load which exceeds the MAX218's supply capability. The TPS7248 micropower LDO (low drop-out) voltage regulator eliminates this problem with a PMOS pass element.

To run the photometer application, first place the filter cap (included with TSL230EVM kit) firmly over the TSL230 IC. See *Figure 9-1* for the evaluation board's outline and major components. Ensure the cap is centered over the square chip inside the TSL230 by sighting through the hole in the cap. The filter is a commercially available color compensating filter (Hoya Optics CM-500) which blocks IR light above about l = 700nm. This corrects the TSL230 sensor's spectral response to approximate that of the human eye. This type of response, also know as the photopic response, is useful in applications such as camera exposure control, lighting control, and display contrast control, which require a measurement of the perceived brightness of ambient light. The software converts the output frequency to another set of unit symbols, such as Lumens/cm^2 or mW/cm^2, to approximate a commercially available photometer.

A Discrete TSL230 Circuit

If you want to experiment with this IC but don't want to purchase the evaluation board, you can do so with the circuit in *Figure 9-9*. Normally, a photometer is an instrument which measures visible light quantity (flux) and brightness (intensity). This

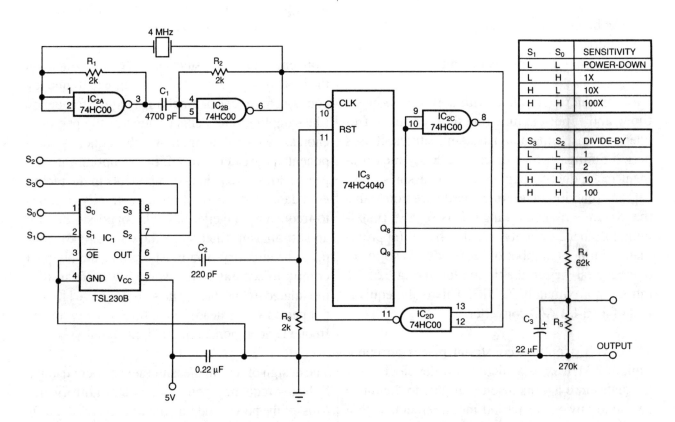

Figure 9-9. A stand-alone TSL230 based light intensity meter.
(Courtesy of EDN Magazine, Chaners Publishing)

application simulates the operation of a simple hand-held-type photometer and displays the output in several ways. However, this circuit uses the TSL230B (5% output frequency tolerance), a 12-bit counter, and a stable crystal oscillator to measure light intensity and display the results on your DVM or a DMM. A 1 volt output equals an irradiance of 1 mV/cm^2.

The 4 MHz oscillator, composed of a crystal and the CMOS NAND gates IC2A and IC2B, form a stable frequency oscillator which clocks the 12-bit counter, IC3, through NAND gate IC2D. The output of the T.I. TSL230B intelligent optoelectronic IC sensor resets the counter through the RC time constant composed of R3 and C2. After IC3 resets, its Q9 output is low, and the output of IC2C is high. IC2D's open gate allows IC3 to count. When the Q9 output goes high, it blocks the gate to disable the clock until IC3 resets again. Each time IC3 resets, the Q8 output of IC3 forms a 32 microsecond pulse. Resistors R4 and R5 divide and capacitor C3 integrates or averages this output to drive a DMM. The fainter the light, the longer in duration the reset time, and the lower the DMM meter reading.

Table 9-1 reveals the test results when you set S_O and S_3 low and S_1 and S_2 high. IC1 has a 1 to 10,000 measurement range with full-scale resolution. The full scale linearity of this circuit, over its entire operating range, is approximately 0.5%. The measurement time is less than 100 microseconds. There is another TSL230 based project (See *Figure 10-15*) which detects and measures UV radiation.

PC SOFTWARE

System Requirements

Real-time operating requirements and the EVM software's graphical nature require a 486/33 MHz or better PC running Windows^{tm} and DOS 3.1 or later.

Installation To install the TSL230EVM evaluation board software:

1. Insert the TSL230EVM software.
2. From Program Manager, choose File, Run.
3. Type A:setup (or B:setup) and press Enter.

After installation, go to the TSL230EVM group and double-click on the TSL230EVM icon to run the program.

From the main screen, there are six buttons, one for each of the applications within the TSL230EVM software. We only investigate the photometer application in depth here.

Software Photometer Operation

When you set the Windows^{tm} software selection to the photometer application, you can access all the functions of the evaluation board by pushing on-screen buttons. You may set the range for manual or automatic gain control. In the manual mode, you press buttons 1, 10, 100, and 1000 for the ranges. When you select each range, the on-screen numerical readout indicates light intensity. If the light intensity is too bright or too dim for each range, the on-screen icon displays an overrange or underrange message. You can select a higher or lower range to prevent the screen displaying overrange or underrange messages.

When you select the automatic range button, the 68HC07 microprocessor on the evaluation board automatically selects the proper range, preventing these overrange and underrange messages. It also sets the light display in the proper readout. If light intensity changes from a completely dark to a bright light setting, the automatic range selects, in sequence, a higher range from 1 to 1,000, as required. As you slowly dim the light, the microprocessor automatically decreases the range setting, corresponding to the dim light. When light instantly changes from bright to dark, the automatic range circuit only drops to the next lower range. It should drop to the lowest range. This problem does not appear if you slowly dim the light.

Light Meter Operation

When you set the software for light meter operation, the screen represents the TSL230 IC in its center. It shows connections to on-screen switches which depict the push-button switches on the evaluation board. In the center of the on screen IC rendition, there is a square, with smaller squares within it. These represent the number of photodiodes you enabled. It also has windows to show the sensitivity, scale factor, and ambient light level. The sensitivity ranges are 1X, 10X, and 100X. The scale factors are 1, 2, 10 and 100. When you point to an icon, switch on the screen and click the mouse actuated button, the switch either moves up or down. This corresponds to the switch position sensitivity and scale factors you changed. One switch controls the S0 function, one controls the S1 function, one controls the S2 function, and the last in this group controls the S3 function. There is another switch controlling the IC's power-down function.

If you press one of the push-buttons on the evaluation board, the corresponding switches on the PC screen change to reflect that action. The settings also include a sample rate selection which may be set to 0.1 second, 0.5 second, or 1.0 second. The light meter reading changes at a faster or slower rate depending on the sample rate setting. The default sample rate of 0.5 second updates the ambient light display every 0.5 seconds.

As the ambient light level varies the numbers on the screen representing its intensity, increase or decrease. If the light is too intense or too subdued for your sensitivity or scale factor settings, an overrange or underrange indicator lights and a screen prompt urges you to change one of the switch settings.

Heart Rate Monitor

When you select the heart rate monitor software, a screen displays a window for power spectrum and another for a plethysmograph (to be explained). There are two knobs controlling a high-pass and a low-pass filter. Another adjustment changes the threshold. This threshold adjustment is a slider pot control icon. The heart rate software only works in subdued lighting conditions, or when you cover the evaluation board with a dark cloth. Place your finger between the high-intensity l = 660 nm red LED, which is part of the TSL235EVM. These are approximately one inch apart. (See *Figure 9-2*)

This demonstration senses light passing through your finger. As it does, blood pulsing through the capillaries modulates the light. Fast Fourier Transform digital filtering techniques recover this low frequency modulation. The resulting waveform is a plethysmograph.

When you press the start button, the evaluation board takes 256 samples at 33.3 Hz, naturally using your default settings. It displays the results in the power spectrum and plethysmograph windows. The power spectrum is a colored bar graph, representing different frequency levels. The plethysmo-

Light source	Distance (ft)	IC_1 output frequency (kHz)	Measured voltage (mV)
25W incandescent	5	1.29	166
75W incandescent	5	5.08	656

Table 9-1. The TSL230's indicated light output frequency vs. distance from sensor and Tungsten incandescent bulb power. (Courtesy of EDN Magazine, Chaners Publishing)

graph window shows a sine wave type graph with varying amplitudes. This graph changes shape, in the manner of an optical low frequency spectrum analyzer, as you change the high-pass or low-pass filter knobs. It also moves in the direction and magnitude of the threshold slider pot.

When you measure your heart rate, hold your finger perfectly still for the most accurate results. The measurement time depends on the number of data points in the sample. The time may vary from 5 seconds to as long as a minute or more. You can adjust these data points and sample rates. You can set the data points for 256, 512, or 1024 samples. You can set the sample rate for 66.7, 33.3, 16.7, or 10 Hz. After completing the samples, a small window in the upper right hand corner of the power spectrum window displays your heart rate.

Proximity Detector Application

Figure 9-10 shows the general principle of operation of this circuit. When you select the proximity detector software, a screen with a window appears. It displays the position of your detected item. There is a display, on the right side of the screen, resembling a vertical panel meter. It has designators of "Near" at the top and "Far" at the bottom. In the center, there is a light which changes color as the needle moves. There is a gain control switch designated "Boost". There is another switch to reject ambient light designated "Ambient".

There are two square buttons labeled START and CALIBRATE. To begin a measurement, press the STARTbutton. Place a reflective object between 3 and 12 inches from the TSL245 optoelectronics

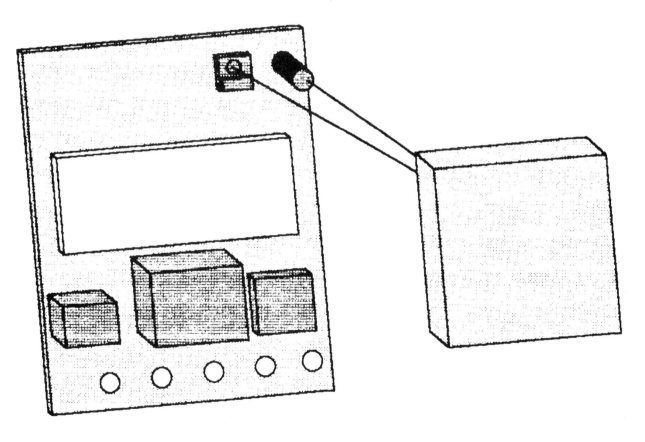

Figure 9-10. *Measuring an object's distance to the TSL235 sensor. (Courtesy of Texas Instruments)*

sensor pair. Press the CALIBRATE button. When you complete your calibration, the meter moves to the center position, and the light turns green. Press the START button again and move the reflective object toward or away from the evaluation board. As the object moves nearer the board, the meter moves upward toward the "Near" position, and the light turns red. As the object moves away from the board, the meter moves downward toward the "Far" position, and the light turns blue. As the meter moves up or down, there is a digital display of the position. It also numerically moves up or down. The meter and the digital position number both represent the proximity of the reflective object to the evaluation board's sensors. The proximity detector perceives the reflective object is either closer or farther away. It does this by comparing the reflected light reading with the calibrated light reading. If the reflected light reading is less than the calibrated value, the object is farther away from the evaluation board. If the reading is greater than the calibrated value, the object is nearer the evaluation board.

Datalogger Application

When you select the datalogger application software, the screen displays a rectangular area with vertical and horizontal bars. This area represents a chart recorder and moves from right to left. There is also a window displaying the sensor's reading. There are buttons to select which applications you use. These buttons have labels of the three on-board optoelectronic sensors:

1. The TSL230.
2. The TSL235.
3. The TSL245.

There are also START and STOP buttons. The last button selects the number of data points you wish to log. This button's default value is 100. You also select the sample rate. The values are: 0.1 second, 0.5 second, 1 second, and 10 seconds. Lastly, you may also select the plot scale. The selection choices are linear and logarithmic.

To begin logging data, select one of the three data sources. The sensor reading should appear in the 100 and 3,000 range. After you set the sample rate and plot scale, press the START button. As samples collect, a yellow line on the chart recorder, moving from right to left, displays them. As the data increases in level, the line moves up. As the data's level decreases, the line moves down. When it collects your selected number of samples, the chart recorder stops. Now you can collect another data sample or save this sample on the display to your computer's hard disk. You can use this method to monitor lighting conditions over a long period.

Chapter 9

Chapter 9 Quiz

1. Which is an IR LED pair?
   A. The TSL235.
   B. The TSL245.
   C. The TSL230.
   D. None of the above.
2. In which application can you click between Lux and footcandles?
   A. The photometer.
   B. The stand-alone TSL230 application.
   C. The heart rate monitor experiment.
   D. The proximity application.
3. Which applications helps over the adverse effects of fluorescent lamp's flicker?
   A. The ambient light synchronization experiment.
   B. The photometer experiment.
   C. The heart rate monitor experiment.
   D. The datalogger experiment.
4. Light energy on a receiving element / the sensor amplifier's threshold, defines what parameter or concept?
   A. Threshold switching in optoelectronic devices.
   B. The device's saturation.
   C. The device's excess gain.
   D. None of the above.
5. What is the purpose of the O/E pin on the TSL230?
   A. It places this optoelectronics device in a tri-state or high impedance mode.
   B. It disables all its functions.
   C. Both A and B.
   D. None of the above.
6. Mode 1 on the TSL230EVM board serves what purpose?
   A. Triggering the stand-by state.
   B. Making radiometric measurements.
   C. Both A and B.
   D. General light measurement applications.
7. What is the major system constraint(s) on the TSL230EVM?
   A. Must draw less than a picoamp of current.
   B. Runs on two 1.5 volt batteries.
   C. Must have a minimum component count.
   D. Both B and C.
8. The stand-alone TSL230 application is what?
   A. A radiometer.
   B. A photometer.
   C. A heart rate monitor.
   D. An indirect means of a blood pressure monitor.

171

Chapter 10
Optoelectronic Projects

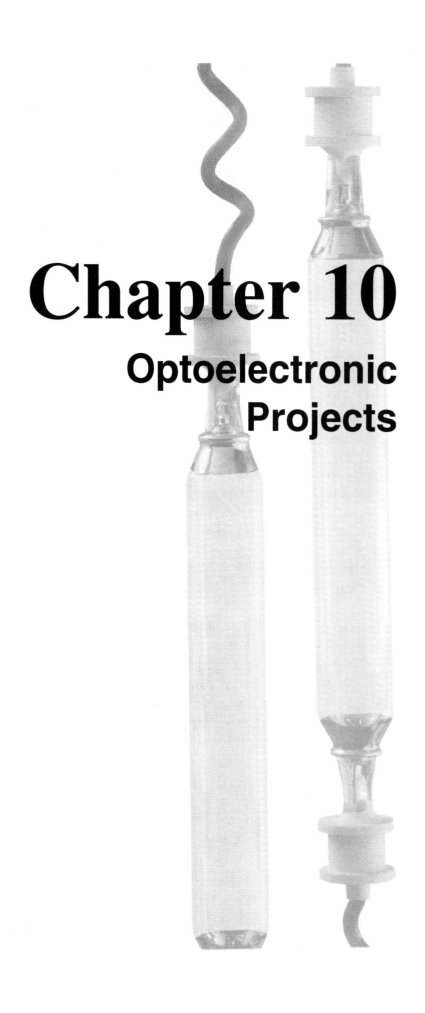

Chapter 10
Optoelectronic Projects

This chapter presents five optoelectronic based projects:

1. A color temperature meter with an extensive tutorial on its underlying theory and related photographic application hints.
2. A UV radiation detector.
3. A sophisticated light-activated home security system.
4. A DTMF (Dual Tone Multiple Frequency) IR LED based car alarm.
5. A binary-to-octal or hex code converter.

To help make these PC boards, there's a non-photographic technique to take 1:1 scale foil patterns off these pages. There is also a negative registration (alignment) project for double-sided PC boards.

THE COLOR TEMPERATURE METER

Photographers use color temperature meters to ensure true representation of colors, especially skin tones. Don't confuse this with a light meter which just measures the intensity of visible light without indicating where it resides within the visible light spectrum. This inexpensive and easy-to-build project, once calibrated, allows you to take consistent repeatable color prints or slides.

The color temperature convention exists because your eye compensates so well for variations in ordinary light's color balance the brain overcompensates or averages out color variances until we believe none exist. Color photographic film reproduces all variances exactly. Improperly matched light sources and film types yield improper colors.

Color temperature reflects the changes in color of a heated piece of metal. As metal is heated, it changes from a glowing red to an orange, followed by a yellow, blue, and finally it reaches a "white" hot state. This convention is used for assigning colored light a color temperature. However, color temperature is based only on the visual appearance of a light source, not on its thermal temperature. For example, the night sky has a blazing color temperature of over 10,000 Kelvin, yet it is below freezing.

The Anatomy of the Human Eye

The lens of the eye is adjustable, like a camera lens. It focuses light images on the retina at the back of the eye. The pupil is also adjustable to determine the brightness or intensity of the light reaching the retina. These adjustable qualities make the eye very "forgiving" of ambient light colors.

The aperture or hole of the pupil is small when viewing brighter light images and larger for viewing dimmer images. This aperture allows the average intensity at the retina to be held constant over approximately a 16:1 ratio variation in the brightness of the object being viewed.

The retina has approximately 100 million rod-shaped cells and seven million cone-shaped cells. The rods respond to dim light (below 0.001 candela per square meter) and the cones require brighter levels (approximately 10 candelas per square meter). Cones are abundant near the center of the retina (the fovea) and detect color. Cones can also distinguish between very small objects. The rods do not sense color.

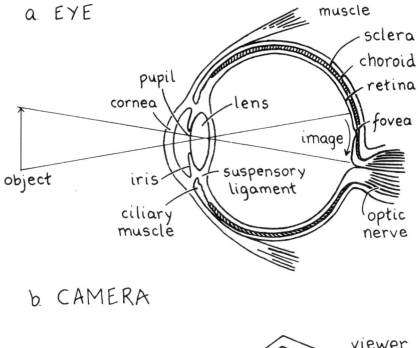

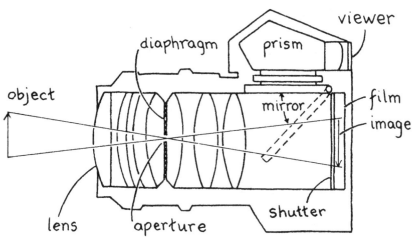

Figure 10-1. The human eye - camera analogy.

Color Perception Within the Eye

A person who is not color blind has, within the retina, color sensitive nerve cells called cones. The eye has no yellow sensitive cones, only red and green sensitive cones. (See *Figure 10-1*) When the eye sees both red and green it interprets this as yellow. This project emulates the functions of the eye involved in perception of color.

Color Perception Within the Brain

The brain has two separate regions for color perception, the chromatic and achromatic. In most instances, both regions help process visual data. However, black and white stimuli disable the chromatic region from processing data, especially complex visual information.

The Ergonomics of Color

The perception of color results from reacting to "visible" wavelengths of electromagnetic radiation. The optical and sensory mechanics of the eye give color its three basic qualities. These are hue, saturation, and lightness. Hue identifies the color in relation to other colors within the light spectrum. Saturation defines the "purity" of a color. As colors within the spectrum become less pure, they start to appear gray or white. Lightness refers to the relative strength of the light coming from a color, as perceived by the viewer. Increasing lightness fades pure colors until they are almost entirely "washed out."

Color Distribution and Saturation

Colors widely separated in the spectrum, such as red and green, are easier to discriminate than neighboring colors. "Grayish" colors of low saturation become difficult for the brain to distinguish. These two phenomena are the basis of our project. Conversely, colors with high saturation, which are also widely separated in hue, require your eye to refocus. This can be a source of fatigue. Your eye's foveal region, which yields maximum visual resolution, is essentially "blind" to blue. This makes blue a poor choice for presenting detailed information.

A Short History of Color

In 1666, Sir Isaac Newton correctly declared that all colors exist in white light. He devised a "color circle" placing red and violet on the ends, bracketing the remaining primary colors in their proper order. A "color solid" set of triangles was next made by the German mathematician Tobias Mayer in 1758. These triangles contained pure red, yellow, and blue on their angles. His greatest contribution, though, was noting that mixtures of pigments are subtractive; therefore, his base triangle had black (the absence of all colors) as its center. He also discovered that there are additive primary colors of blue, green, and red light. These three colors are called "primary" because all other colors can be made by adding these three together in various intensities.

In 1810 a German classical painter, Philip Runge, wrote *Die Farbenkugel* (*The Color Globe*) in which he placed pure colors along an equator and had gradations of a primary red, yellow and blue. In 1876 Wilhelm von Bezold devised a cone with white at its base and black at its tip and all the visible spectrum in between, something similar to Mayer's work. But the most definite work was by the 1909 chemistry Nobel Prize winner Wilhelm Ostwald. His 1916 work, *Die Farbenfibel* (*The Color Primer*), divided colors into two classes:

1. The achromatic colors white, gray, black and shades in-between.
2. The chromatic colors red, yellow, blue, green and all colors adjacent to them.

The commercial world, more than the artistic world, recognized Ostwald with a sponsorship from the Container Corporation of America. When Hitler assumed power, one year after Ostwald's death, he had all of Ostwald's theories taught to German school children. In fact, Hitler went beyond Ostwald's equal value scales and developed even more elaborate scales of his own. A "value" is the darkness or lightness of a color; brightness is the common, less technical way, of stating this.

America's color theorist was the painter Albert H. Munsell (1858-1918), whose work, *A Grammar of Color*, proposed the concepts of hue, saturation and brightness, as we explain in this chapter. In *A Grammar of Color*, Munsell developed a "color tree" with a trunk as a thin column that shows the scale of brightness, ranging from black to white. The hues of the visible spectrum are arranged around the trunk. Each branch shows the chroma from very weak at the trunk to very strong at the end. Munsell assigned a number to each spot on his tree; therefore, it is possible to express a color exactly by a number.

Unintended Discoveries and Their Consequences

"Man is a shrewd inventor, and is ever taking the hint of a new machine from his own structure, adapting some secret of his own **anatomy** in iron, wood, and leather, to some required function in the work of the world."

Ralph Waldo Emerson (1803-1882)
from *English Traits*, "Wealth"

A Unique Use of the Camera's Direct Predecessor

If necessity is the mother of invention, there must be a prankster somewhere in this grand scheme to account for the large number of accidental discoveries of ultimately great value. One such case is the creation of a new field in science resulting from one of the first uses of your modern camera's direct predecessor. In 1839, Louis Daguerre demonstrated his "camera obscura". Assisting artists to draw more lifelike scenes with more exact scales was the necessity of this invention. It consisted of a box fitted with a lens at one end and a slanted mirror at the other. The lens condensed or "miniaturized" the mirror's reflected image, casting it towards the top of the box upon a planar ground glass screen (often with grids). Placing onion skin like translucent paper over the ground glass grid permitted artists to study the scene and more accurately trace it. Today's camera evolved primarily from the "camera obscura" by replacing its tracing paper with light sensitive film.

Accident is Now The Mother of Invention

An English gentleman within the prim and proper Elizabethan age became fascinated with England's "criminal" element. He knew his mere empirical observations though were not scientifically based. He therefore sought to unearth some common thread permeating those predisposed to crime. He commissioned a portrait artist to draw portrait sketches of a great number of criminals. The artist used a "camera obscura" and sketched their faces using the ground glass screen's grid for uniform scales and to ensure approximate correct placement on the sheets. He did not know just how precise these translucent sheet portraits were in relation to one another. The Englishman soon discovered this quirk of good fortune, but quite by accident! As he gathered and neatly stacked these criminal portraits, to place in his attache case, he glanced down and saw the astounding "average" of these composite overlays. He then reconfirmed this discovery by holding them against a strong light source.

He tried rearranging the stack in a myriad of random combinations, and the same startling result persisted. Viewing them in any order always disclosed a far more appealing "composite" man. This demonstrated criminals with faces lacking reasonable symmetry were less appealing; but, by averaging multiple people's features, it smoothed out irregularities. An extreme example of asymmetry is the Hunchback of Note Dame. This likable cartoon character has, among other unique features, hideously asymmetrical eyes.

Types of Color Film

There are three main types of color film: Daylight, Tungsten and Type A. Daylight film is balanced for the heavy color temperatures occurring near noon from electronic flashes and blue flash bulbs with a 5500 Kelvin color temperature. Tungsten film is balanced for incandescent lamps with a 3200 Kelvin color temperature. Most indoor films are Tungsten balanced. Type A film is especially color balanced for use with photo floodlamps, with a color temperature of 3400 Kelvin..

Colored Film/Light Source Incompatibility

A color photo taken inside under a filament light, using ordinary "daylight" film, will come out with an assortment of annoying reds, oranges, and yellows. Greens and blues will be almost nonexistent. Areas that are white, such as a shirt, will turn out yellow. The opposite effect occurs when taking photos underwater. Blues and greens will dominate; reds and yellows will appear almost white because suspended particles in natural water absorb red wavelengths.

The most accurate color rendition is obtained with the exact match of the film with the color temperature of the illumination. The color temperature is marked on most electronic flashes and on photographic bulbs. *Table 10-1* lists color correcting filters and the following section on color filters explains their theory.

A Primer on Color Filters

These come in two classes: correcting and light balancing. Correcting filters make large shifts in color temperature. Light balancing filters make smaller more subtle changes in color temperature. Correcting filters allow you to use daylight film and expose it under tungsten light sources and vise-versa. These filters come in two series.

The 85 Series of Correcting Filters

These are amber "warming" filters. They absorb blue light and convert it to color temperatures of various tungsten light sources.

The 80 Series of Correcting Filters

These are blue "cooling" filters. They absorb yellow light, converting film designed for tungsten light to be used outdoors under daylight. Incidentally, daylight has an assigned color temperature of 5500 K to 6500 K. This is slightly subjective since daylight is a mixture of sunlight and skylight (10,000 K), yet both appear white to our eyes. This is caused by the eye's forgiving nature. An 80 series filter can shift a zirconium filled flashbulb (80D) or an aluminum filled flashbulb (80C) up to daylight color temperature for use with daylight color films.

Light Balancing Filters

These make smaller shifts in color temperature, e.g., changing 3200 K lights for use with 3400 K tungsten film. These subtle filters come in the bluish 82 series and the yellowish 81 series. (See *Table 10-1* again)

The Mired System

The two types of filters previously mentioned make changes in color temperatures of light sources. However, these shifts are of a certain magnitude at a predefined color temperature. The mired system allows a certain magnitude shift regardless of what the color temperature is. The mired system is an acronym for (Micro-Reciprocal Degrees) and is the color temperature divided into one million. As an example, a 3200 K Tungsten light source is 1,000,000 divided by 3200 for a mired value of 312. This number is then is divided by 10 and called a *decamired* value of 31 (312 divided by 10). Note the rounding off of .2, the least significant digit.

TYPE OF FILM USED	EXPOSED BY:	FILTER TO USE	OR ELSE:
Daylight	3200 K	80A	too orange
Daylight	3400 K	80B	too orange
Daylight	Aluminum filled flashbulbs	80C	too orange
Daylight	Zirconium filled flashbulbs	80D	too orange
3200 K	Daylight	85D	too bluish
3400 K	Daylight	85	too bluish

LIGHT BALANCING FILTERS

FILTER USED:	ALLOWS LIGHT AT THE TEMP. IN THIS COLUMN TO BE USED WITH 3200 K COLOR FILM:	ALLOWS LIGHT AT THE TEMP. IN THIS COLUMN TO BE USED WITH 3400 K COLOR FILM:
81	3300 K	3510 K
81A	3400 K	3630 K
81B	3500 K	3740 K
81C	3600 K	3850 K
81D	3700 K	3970 K
82	3100 K	3290 K
82A	3000 K	3180 K
82B	2900 K	3060 K
82C	2800 K	2950 K

COLOR CORRECTING FILTER TRANSMISSION/DENSITY DATA

DENSITY	TRANSMISSION	ABSORPTION
0.05	90%	10%
0.10	80%	20%
0.20	63%	37%
0.30	50%	50%
0.40	40%	60%
0.50	30%	70%
0.60	25%	75%
0.70	20%	80%
0.80	16%	84%
0.90	12.5%	87.5%

QUALITATIVE DATA ON COLOR ABSORPTION & TRANSMISSION

A FILTER OF THIS COLOR:	WILL ABSORB:	AND TRANSMIT THIS:
Red	Cyan (Blue & Green)	Red
Blue	Yellow (Red & Green)	Blue
Green	Magenta (Red & Blue)	Green
Magenta	Green	Magenta
Cyan	Red	Cyan
Yellow	Blue	Yellow

Table 10-1. A summary of color correction and conversion filters.

Using the Mired System

Suppose you had daylight film with a color temperature of 5500 K, that would be 1,000,000 divided by 5500 for a mired value of 181.8 or a decamired value of 18. However, you want to use the film indoors under a Tungsten lamp with a color temperature of 3200 K (312.5 or 31 decamired value). The difference in color shift is 31 minus 18 or 13 decamireds. You could use two mired system filters, 12 and 1, which add for a total of 13. These could be a B-1 and B-12 filter with "B" standing for the bluish cooling color.

Let's assume a manufacturer of film made an absurd film with a color temperature of 16,000 K and you want to shoot in a daylight setting of (5,200 K). The two mired values are 62.5 and 192.3 respectively. This is a difference of 129.8 or a decamired value of 13. You could use the same combination of filters we used in the preceding example with equal effect. Obviously no 16,000 K :exists, this was only used to demonstrate that the color shift is a constant amount and not dependent on where you are when you start the color temperature shift correction.

CC Filters

Color compensating (CC) filters are the last type of optical filters and come in at least six colors. They come in a variety of primary colors so you can control one or two of the additive primaries, leaving the others unaltered. As an example, a CC20Y is a color correcting yellow filter with a 0.20 density to blue. That means it absorbs 37% of the blue light. A CC50R is red and has a density of 0.50 to blue and green (it absorbs about 68% of blue and green light). (See *Table 10-1*)

The Kelvin Temperature Scale

William Thomson (Lord Kelvin), developed a universal thermodynamic temperature scale in the early 1800s based on the coefficient of expansion of an ideal gas. Kelvin also established the con-

Color Temperatures of Various Objects	
Candle Flame	1500 Kelvin
60 Watt Filament Bulb	2800 Kelvin
250 Watt Photoflood Lamp	3400 Kelvin
Sunlight at noon with a blue sky and few clouds	6000 Kelvin
Blue Sky	10,000 Kelvin
Blue Northerly Sky	20,000 Kelvin

Table 10-2. Color temperatures of various familiar objects.

cept of absolute zero, the temperature (not yet attained) at which all molecular motion ceases. Absolute zero equals -273.15°C. Kelvin is the Celsius temperature with 273.15 added to it. It is expressed in the following equation:

$$K = \mu \,°C + 273.15$$

In more practical terms, a 100 Watt Tungsten filament bulb has a color temperature of 2850 K. Refer to *Table 10-2* for the color temperature of some familiar objects. The light from a similar bulb of 40 Watts is only 2750 K since it dissipates less power and burns at a slightly decreased temperature. The lower the color temperature, the more closely the object resembles red (the lower end of the light spectrum). As the object heats, the color temperature increases, approaching the violet end of the spectrum.

The Blackbody Concept

This is a device to which all irradiance measurements are referenced. Theoretically, a blackbody is a perfect radiator and absorber of radiant energy. Its radiation spectrum is therefore a simple function of its temperature. This is why mercury vapor and fluorescent lights are not usable with this color temperature concept. Fluorescent lights discriminate against the emission of certain bands of color temperatures or energy. Fluorescent tube lamps hold argon and mercury gases which are excited by electric current, producing ultraviolet light. This UV light strikes the tube's inner phosphor coating, creating a glow and emitting various lights all

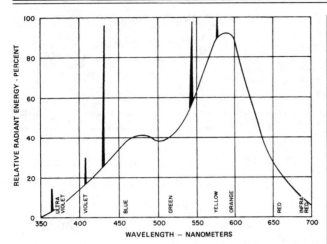

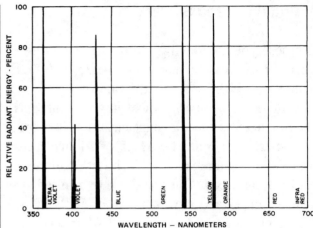

Figure 10-3. The spectral color content of a mercury vapor bulb.

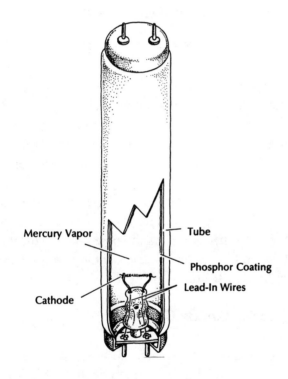

Figure 10-2. A fluorescent bulb's construction & spectrum.

with different peaks within the visible spectrum and beyond. Light output occurs in bands (See *Figure 10-2*) not a continuous single color, such as a blackbody produces. Mercury vapor lights, often used as yard security lamps, also exhibit a similar multiple peak spectral signature. (See *Figure 10-3*) Compare these to a Tungsten lamp's very uniform spectral output. (See *Figure 10-4*) A visible

LED, however, has a narrow single band color spectrum.

There are special purpose fluorescent lights made by Philips, the Ultralume™ series, which emits only one band of radiation. As an example, their 50U has a color temperature of 5,000 K. Places where the colors of objects on display is important, such as museums and jewelry stores, use these special bulbs.

Photographic Conventions Are Hard to Shed

In some cases photographers purposely seek unbalanced lighting conditions. Light is reddish late in the day because the low lying sun filters the earth's atmosphere to a color temperature of about

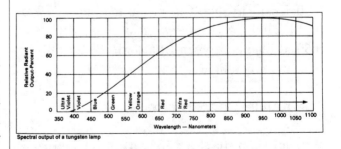

Figure 10-4. The spectral color content of a tungsten bulb.

Chapter 10

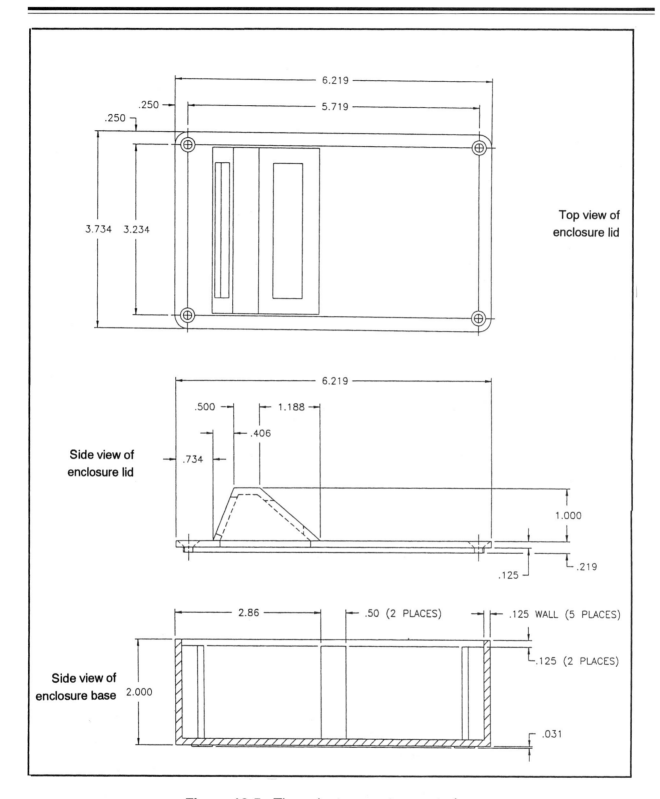

Figure 10-5. The color temperature meter's case.

5500 Kelvins. Often, photographers shoot this light unfiltered for the dramatic effect of the reddish cast that looks like a sunset. In a color photo of a field of snow with a dark shadow, the shadow area comes out illuminated, mainly from the clear blue skylight of a very high color temperature. The shadowed snow comes out blue when photographed with ordinary daylight film. A pale "skylight" fil-

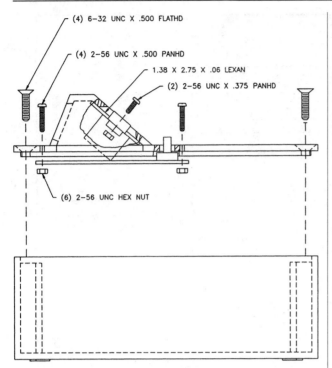

Figure 10-6. Attaching the PC board to the case's upper lid.

crumpling provides diffusion of the incoming light. You compare this light against a dual color LED (red and green) which emits varying colors of light housed in a segregated adjacent chamber. (See *Figure 10-5*) Maintaining the light separation and integrity of these two chambers is essential to this project's success. *Figure 10-6* shows how to mount the PC board to the case's upper lid.

Advancing the momentary contact push-button switch on the front panel in 16 equal increments changes the color of the LED from totally red to totally green while resembling all of the yellow and orange hues in between. When the color of the dual LED matches that of the incoming light, note the meter position. Depending on how far it has deflected, this is the color temperature of the incoming light. After calibrating the instrument against some well known sources (as we will do later), you'll be able to accurately equate the meter's deflection with color temperature.

The Schematic Diagram

Figure 10-7 is a block diagram of the color temperature meter. *Figure 10-8* is the schematic. The two "D" flip-flops and ICs U1 and U2, form a unique 4-bit counter. This unique counter also serves as a switch debouncer, which eliminates the need for superfluous cross coupled NAND gates with pull-up resistors. The dual op amp, U3, is both an inverting summer (U3A) and an inverting amplifier (U3B).

ter, designed to block some of the light from the sky, would reduce the bluish cast of the shadows.

How it Works

The project's case is uniquely constructed with two light-tight chambers. One allows unimpeded light to enter from the outside. This light is "warmed" or color temperature shifted by a yellow optical filter or crumpled yellow cellophane wrap. The

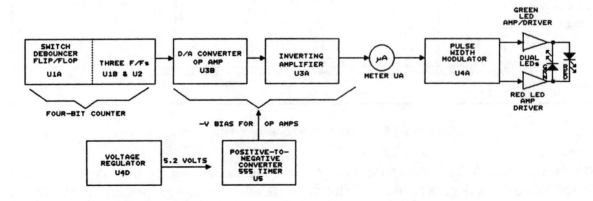

Figure 10-7. The color temperature meter's block diagram.

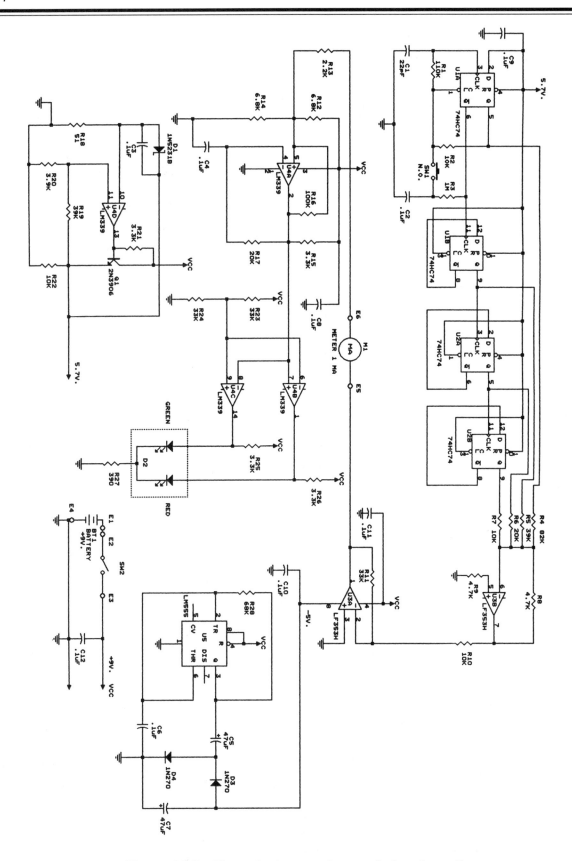

Figure 10-8. The color temperature meter's schematic.

LED Duty Cycle Vs. Color	
Percentage ON (Duty Cycle)	Resultant Color of the Dual LED
67%	Green
56%	Green/Yellow
50%	Yellow
43%	Orange
34%	Red

Table 10-3. LED duty cycle versus color.

The LM339 Quad Open Collector Op Amp

The LM339 op amp, U4, has open collectors; note the pull-up resistors used on all its outputs. Any resistor in the 2 to 5 kilohm range is acceptable. The LM339 quad op amp, U4, serves three purposes. First, U4A is a voltage controlled Pulse Width Modulator (PWM). Second, amplifiers U4B and U4C comprise a current source which drives the dual color LED. This inverting/non-inverting pair amplifies a varying duty cycle from approximately 30 to 70 percent. (See *Table 10-3*) Duty cycle depends on the control voltage and the biasing network of R12, R13, and R14. The amplifier pair's varying duty cycle drives the red-green dual LED with varying amounts of current. The LEDs' color is proportional to duty cycle. (See *Table 10-3*) Third, U4D, is a voltage regulator. It produces a regulated 5.3 volts. This is ideal for the 74HC TTL family of logic (ICs U1 and U2). These ICs operate over a 2 to 6 VDC range.

The 555 Voltage Converter

The last IC, U5, is a 555 timer configured as an astable or free running multivibrator. The 555 drives the capacitor/diode pairs of D3/D4 and C5/C7. This produces a negative 5 volts out; but it suffers from poor regulation. That is why we lightly load it by supplying the negative bias to U3. Note U3 is a BIMOSFET amplifier, which draws minimal current.

A Unique Counter Circuit

Switch SW1 clocks this unique 4-bit counter and switch debouncer. Integrated circuit U1A acts like a true/complement buffer. Resistor R1 and capacitor C1 ensure IC U1A comes out of reset prior to the clock's edge. Resistor R2 applies IC U1A's logic state to pins 1 and 3. When the switch closes, the next logic state (which is stored on the capacitor) is transferred to the flip-flop's reset and clock inputs. Releasing the switch (remember it is a spring loaded N.O. keyboard type switch) allows the capacitor to charge to the next logic state through R3. The counter advances states as you first press the switch, rather than when you release the switch.

Counter Circuit Modifications

You may replace capacitor C1 by a single 3 megohm resistor. If you don't use C1, make R1 and R2 each ten times larger than the schematic values. Using a capacitor for C1 instead of a resistor reduces the circuit's sensitivity to parasitic effects. Also, resistors above 1 megohm are often hard to obtain. The counter purposely starts at a state of 0100 or decimal 4 at power turn-on. This deflects the color temperature meter to one-fourth scale,

Oscillator Trip Points & Taken Data			
Step No.	Voltage at U3B-7	Duty Cycles of Pulses to the Dual LEDs	
Step 1	.5625 Volts	24.2%	22.7%
Step 2	1.125 Volts	28.2%	29.7%
Step 3	1.6875 Volts	35.5%	33.8%
Step 4	2.25 Volts	39.0%	38.3%
Step 5	2.8125 Volts	43.6%	41.6%
Step 6	3.375 Volts	45.0%	46.6%
Step 7	3.975 Volts	49.2%	49.3%
Step 8	4.500 Volts	52.8%	53.4%
Step 9	5.0625 Volts	56.9%	55.3%
Step 10	5.625 Volts	60.3%	60.5%
Step 11	6.1875 Volts	65.5%	64.7%
Step 12	6.75 Volts	67.5%	69.4%
Step 13	7.3125 Volts	70.0%	73.6%
Step 14	7.875 Volts	73.9%	76.4%
Step 15	8.4375 Volts	77.0%	84.9%

NOTE: The first duty cycle column is with R13 = 2.2K, and R12 and R14 both = 6.8K. The second column is with R13 = 2.2K, and R12 and R14 both = 7.2K.

Table 10-4. Oscillator trip points and data taken.

indicating that both the instrument is powered up, and that you have a good battery.

The Inverting Summer

Each bit of the 4-bit counter is weighted or ascends in a doubling of the resistor's values from the MSB to the LSB. The resistors in the summer's input are nearly double one another. They are as close as standard 5% value resistors will allow. The doubling effect of the resistors makes the inverting amplifier reflect the counter's weighted inputs. They also ascend in a binary fashion.

The inverting input node sums these currents and the feedback resistor, R8, determines the circuit's gain by the following equation:

$$V_{OUTPUT} = -R8 * [(V/R4) + (V/R5) + (V/R6) + (V/R7)]$$

Where V = a logic high of 5.3 volts from the Q outputs of U1 and U2.

Resistor R9 should reflect the algebraic sum of the parallel combination of the input and feedback resistors. This will minimize voltage offset by helping balance or match the currents flowing into both of op amp U3A's input terminals. Theoretically, no current should flow into either input of an **ideal** op amp. The U3A op amp's output is a negative voltage, representing counts of zero to 15, starting at 4. Op amp U3B inverts this negative voltage and amplifies it with a gain of approximately minus 3.3 or the ratio of 33K to 10K (R11 to R10).

Voltage Inversion for Op Amp Bias

Op amp U3 requires a negative voltage on pin 8. You could use another +9 volt battery, placing it in "backwards" with its +9 volt terminal tied into system ground. However, it is possible to use a single battery by generating a negative voltage from a positive one through the voltage conversion process.

Voltage Conversion Using a 555 Timer

The 555 timer, U5, provides a continuous stream of square waves occurring at a frequency of 10 kHz. This higher frequency helps minimize the values of storage capacitors C5 and C7. These capacitors, along with rectifier diodes D3 and D4, give an approximate negative 5 volts output. To further enhance efficiency, use capacitors with low ESR[1] ratings for C5 and C7. The rectifier diodes should have as low of a voltage drop as possible to minimize losses. Use germanium diodes or similar silicon small signal diodes such as 1N270, rather than 1N914.

Circuit Precautions

Supplying the 555 timer with +10 to +15 volts optimizes this circuit's operation. But wait a minute, we only have +9 volts from our battery. That is true, and if it proves to be a problem. This allows you to either use a voltage doubler scheme or to increase the gain in U3B. This should not be necessary though unless the U3 op amp you use is a bipolar one which draws more than 2 mA.

The Metering Circuit

This consists of a 1 µA meter. Ideally, you can use a small inexpensive edgewise meter. These are common in tape recorders and other portable consumer electronics products. Meters with a red and green section are best. The red section can be placed at the lower end, corresponding to the occurrence of red at the lower end of the visible spectrum. The meter's green section represents colors toward the other end of the visible spectrum.

Advancing the Meter

Current accumulates and flows into the inverting summer junction of op amp U3A as you advance the 4-bit counter from 0 to 15. This increases voltage by approximately 1/16th times 9 volts per step or SW1 switch depression. The top of the meter's

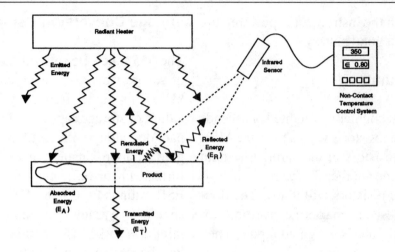

Figure 10-9. Controlling pulse width sensitivity.

scale reflects current from a source which is slightly less than +9 volts. This is due to op amps typically not reaching their "rails" or supply voltages. This control voltage, proportionately represented by the meter's deflection, controls the duty cycle of the PWM, op amp U4A. The control voltage appears at pin 5 of U4.

The Voltage Controlled PWM

The RC time constant (τ) of R17-C4 sets the output frequency. However, this frequency is not critical because a varying duty cycle squarewave strobes the LED at a far greater frequency than the flicker frequency limit of the human eye.

Setting Voltage Sensitivity

The circuit is voltage sensitive due to the divider network R12, R13, and R14. With the input voltage divided down to 50% of the supply voltage, the duty cycle is also about 50%. However, as the control voltage changes, so does the duty cycle. This is due to changing the trip points at which the oscillator functions. (See *Table 10-4*) The upper and lower trip points vary by changing the values of this three resistor voltage divider configuration, especially R13. Different supply voltages also effect duty cycle. Theoretically, the upper and lower triggering points occur at 1/3 and 2/3 of the supply voltage. This is only true if R12 and R14 are equal, and if R16 is not equal to these resistors.

Controlling PWM Sensitivity

Reducing the input to resistor (R13) increases pulse width sensitivity; conversely, increasing R13 with respect to both R12 and R14 reduces sensitivity. (See *Figure 10-9*)

Troubleshooting

If the circuit does not operate after construction, first check the power and ground. There are two voltages within the project. The battery's 9 volts and the voltage regulator's 5.3 volts share a common ground. Follow *Table 10-5* and measure the voltage and ground on each component listed. If that fails, visually inspect each solder joint and

Power & Ground Check-Out Points	
Power	Ground
U1-2, 4, 10, 13, 14	U1-7, U2-7
U2-1, 4, 10, 13, 14	U3-3, U4-12
U3-4	U5-1, C1, C2, R9
U4-3	R14, C4, R18, R20, R22
U5-4, 8	R24, R27, C6, C7+, D4 cathode
R12, R21, Q1c (collector), R23, R25, R26	

Table 10-5. Power and ground check out points.

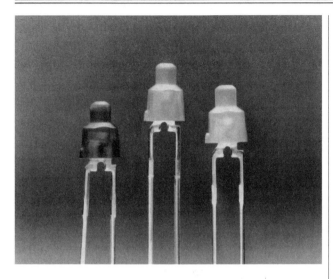

Figure 10-10. *A two lead dual color LED's notched pin (on the left).*

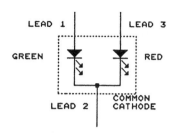

Figure 10-11. *A three pin dual color LED.*

begin tracing the power from the switch through each component listed in *Table 10-5*.

If the colors are reversed in the calibration procedure, reverse the LED. Dual LEDs often have the GREEN anode with a notched or 90° lead. (See *Figure 10-10*) The opposite lead has a 45° "gradually sloped" lead. This circuit is designed for a three-lead dual LED with a common cathode. (See *Figure 10-11*) Due to the circuit's drive scheme, a two-lead dual LED will not work. If you have difficulty obtaining sufficient brightness, use a lighter piece of yellow cellophane for your optical filter or use an enclosure with a larger opening for the incoming light.

Circuit Modifications

Five of the most common problems you might experience are:

1. Insufficient gain from U3A.
2. Insufficient negative bias on U3.
3. Improper meter deflection.
4. Insufficient dual LED brightness.
5. An incorrect voltage out of the U4D voltage regulator.

Insufficient Gain

You can solve this by increasing R11 or decreasing R10, in direct proportion to the desired increase of gain.

Insufficient Negative Op Amp Bias

The circuit in *Figure 10-12* is a voltage doubler and is essentially just adding on another stage to the two diode/capacitor stage already on the 555's output.

Improper Meter Deflection

Reverse the leads if the meter deflects in the wrong direction. If the deflection is not enough, check to see that the meter is a 1 mA movement. If the meter is correctly specified and the deflection is still insufficient, increase R11 or decrease the values of

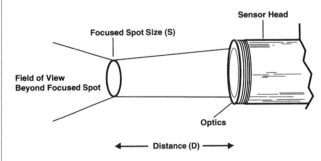

Instrument spot size. Optical resolution of an object is defined as the ratio of the distance from the sensor to the object, compared to the size of spot that is being measured.

Figure 10-12. *A voltage doubler circuit.*

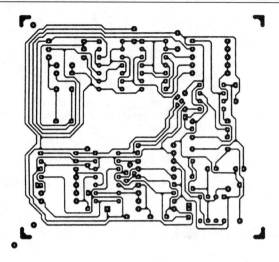

Figure 10-13. The color temperature meter's foil pattern.

R12, R13, and R14 in proportion to one another. If meter deflection is too great, reduce R11 or increase R10. Changing values of R12, R13 and R14 is a last resort since it affects duty cycle and requires recalibration.

There are three basic analog deflection meter movement types:

1. The moving magnet.
2. The taut band.
3. D'Arsonval or Weston movement.

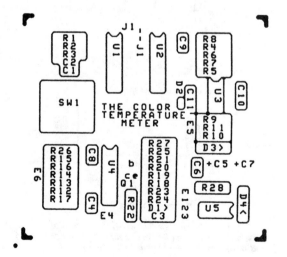

Figure 10-14. The color temperature meter's placement drawing.

The first is the least expensive and has up to 5% accuracy; however, it is subject to the Earth's magnetic field and definitely responds to a magnetic field around an inductor or transformer. This is especially true with lower values such as 100 µA. The second, a pivot jewel or moving coil type meter, is sometimes referred to as a D'Arsonval movement after the man who patented this idea. Edward Weston developed a moving-coil instrument from the delicate laboratory apparatus of D'Arsonval, and realized a ruggedized panel meter. Today the moving coil meter is often called a Weston movement. The third, a taut band uses a stretched band to minimize errors and is the most accurate of these three meter movements. These two more accurate versions of the panel meter may not be available in sizes small enough to fit into the instrument's case.

Insufficient Dual LED Brightness

You can easily solve this by lowering the value of R27. When calculating the LED current, it is on for an average of only half the time. So it is not totally a straightforward calculation using Ohm's Law. You must take the LED duty cycle into consideration in these calculations.

Incorrect Voltage Out of the U4D Voltage Regulator

The voltage regulator in op amp U4D supplies a nominal 5.3 volts to the 74HC series ICs, U1 and U2, and should not go above 5.8 volts to safely remain within the ICs' limits. If problems occur check to see that D1's polarity is correct. If the voltage is still too high, increase R19 or decrease R20 in proportion to the amount you wish to reduce the voltage. Change these resistors in the opposite direction for too low of an output voltage. This resistor value change should typically be no more than within the 5 to 10% range.

Chapter 10

Construction

Figure 10-13 is the foil pattern of the single sided PC board. *Figure 10-14* is its component placement drawing. The board has a single jumper, indicated by a dashed line.

Construction Procedure

The interior of the case should be matte black plastic. Use the case specified, or spray paint your own so no internal reflections occur. The case in the parts list has a raised portion suitable for displays. This makes it ideal for this project. (See *Figure 10-5*) Maintaining light integrity of the two adjacent chambers is crucial when constructing this project. Light containment prevents the larger chamber, designed for incoming light, from "bleeding" over into the chamber housing the dual LED. This would degrade LED brightness, but more importantly, it would also distort color temperature comparisons.

The case has to be modified slightly. As you hold the case, the side of the display facing away from you has to have an aperture cut in it. This allows light to enter. The front panel also has to be modified. You need holes for the push-button switch to push through, plus mounting holes for the PC board. You need standoffs to hold the board a half inch away from the front panel.

The filter may be a glass or gelatin type optical filter. The aperture should be as wide as possible to allow operation in low light conditions. You may want to have a sliding shutter to allow less light to shine in during high ambient lighted conditions.

By using the table of light sources you can count the number of push-button switch advances and fill in the 16 entries, creating your own calibration table. *Table 10-6* reflects results this author obtained. Once you have calibrated your meter and filled in the 16 entries you will know that, for instance, a quarter scale deflection means 4,500 K and half scale means 6,000 K, etc. It may vary, depending on the dual colored LED you use.

There is a relationship between Lumens/Watt and color temperature. If you know one you can approximate the other. This corresponds to about 2900 K in *Table 10-7*. This table shows how you can convert a projector lamp's Lumens per Watt to

Pushbutton Switch Position Vs. Color Temperature	
Switch Depression	Resultant Color
1	Red
2	Red
3	Red
4	Reddish-Orange
5	Reddish-Orange
6	Reddish-Orange
7	Orangish-Yellow
8	Orangish-Yellow
9	Orangish-Yellow
10	Light Yellow
11	Darker Yellow
12	Yellow
13	Yellowish-Green
14	Green
15	Darker Green
1	Recycles through the colors*
*If your colors are reversed from this pattern, you've placed your LED in backwards.	

Table 10-6. Push-button switch position versus color temperature.

Lumens Per Watt to Kelvin Conversion	
Lumens/Watt	Kelvin Temperature
4	2175
8	2460
12	2675
16	2850
20*	3000
24*	3125
28*	3250
32	3350
36	3500
* These are approximate ranges of projector lamps.	

Table 10-7. Lumens per Watt to Kelvin conversion.

color temperature in Kelvin for the lower end of the visible light spectrum. To obtain this approximation of Lumens per Watt, just divide the Lumen rating by the number of Watts. As an example, a 100 Watt frosted light bulb is rated at 1740 Lumens or 17.4 Lumens per Watt.

Assembly

After mounting the components onto the PC board, check the functions by advancing the meter and watching the LED change colors. The final 11 step assembly of the color temperature meter is:

1. Connect the two jumpers (J1s) together with a single wire approximately .4" long.
2. Solder all of the components, the switch, and battery wires to the appropriate E numbers.
3. Mount the PC board (steps 4 and 5).
4. First, drill two holes in the front panel. (See *Figure 10-5*)
5. With the lid removed, install the PC board.
6. Cut and place the piece of mylar in the front opening.
7. Cut and place the yellow cellophane in the rear opening.
8. Place the meter in the space to the left of the dual LED and the mylar front panel. Wire the meter in.
9. Secure the meter with a metal bracket commonly supplied with the meter. If no bracket exists, glue it in with a quick drying adhesive.
10. Place a battery into the holder and turn on the switch.
11. Make sure the meter deflects approximately one-fourth scale.

If this does not occur, see the Troubleshooting section; otherwise, proceed to the calibration section.

Calibration

Calibrate this project by comparing objects with a known color temperature spanning the visible spectrum. See *Tables 10-2* and *10-7* for these lists. The last three references in *Table 10-5* are subject to a slight degree of error due to their dependence on location and variations in atmospheric conditions. When calibrating this instrument, shade it from the sun and don't have objects inadvertently cast light directly into the instrument. Examples include a tree with green leaves, a bright red brick building, or a person who is wearing bright blue clothes. It is best to calibrate the instrument in total darkness unless you have thick curtains in your lab or workshop.

THE UV RADIATION MONITOR

Figure 10-15 detects the intensity and presence of 295 *nm* radiation. This is right in the middle of the harmful UV-B band. It is based on a Texas Instruments TSL230 light-to-frequency IC. To start this easy-to-build, battery operated project, press switch S1 for at least 2.5 seconds. This start pulse begins the 555 timer's 2.2 microsecond measuring cycle. Pin 3 of the 555 timer resets the 4040 CMOS binary counter. UV radiation at 295 nm produces a current in the Hamamatsu vacuum photodiode, P/N R1107. This photodiode has a 295 nm UV passing filter in front of it. The photo induced current charges capacitor C1 which is from 5 to 20 pF. Adjust capacitor C1 to give the proper span from all ten LEDs on at noon on a bright summer day, to just one LED on when inside with no UV radiation, such as indoors at night.

The two op amps, IC1 and IC2, are non-inverting amplifiers having a gain of approximately 20. The amplified linearly increasing voltage triggers the second 555 timer, IC3. You could use a dual 555 timer, the 556, instead of two timers. If the ramp voltage reaches two-thirds of timer IC3's supply voltage, an output pulse is applied to the clock (pin 10) of the 4040 CMOS binary counter, IC5. Simultaneously, this discharges capacitor C1 through the second timer, IC3's pin 7. This cycle repeats until the first 555 timer, IC4, times out and its pin 3 goes to a logic low. This IC4 is the timer which actually holds the radiation count and drives the LED bar graph 10-LED array. If the UV radiation

Chapter 10

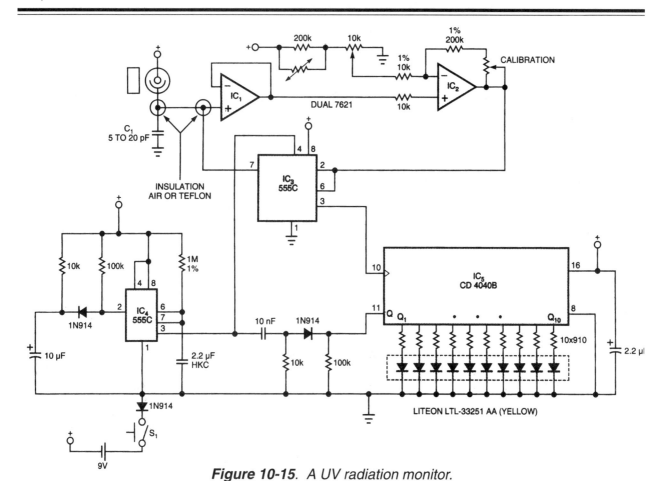

*Figure 10-15. A UV radiation monitor.
(Courtesy of EDN Magazine, Cahners Publishing)*

sensor detects no UV, bit 1 still lights, indicating the battery is operative and possesses sufficient charge.

THE BURGLAR BAFFLER PROJECT

The conventional light timer is one of the most economical and practical burglar deterrents. However, in its current state, it has three severe limitations which the Burglar Baffler overcomes. The first and most objectionable limitation is the predictability of the lights. They are turned on and off at the same time rather than having a multiple and varied number of light activations within an hour or 24-hour period. The second problem is the limited life and cost of frequently replacing the conventional light timer, due to its mechanical nature. For example, the current carrying contacts soon become pitted, undependable, and therefore largely ineffective. The third limitation is a subtle problem often overlooked. This comes from not protecting the incandescent light bulb filament against current surges which reduce the light bulb's life. This could cause the light bulb to fail while you are away.

The Burglar Baffler is programmable over several days, adds pseudo-random actuations, and allows 28-day cycling. You can build and test this project in two evenings for approximately $30. The foil patterns for the printed board are included. This project is designed to be tested with a simple multimeter.

Solving Limited Timer Life

A solid state design and a light activation component, using two 200 VAC series LASCRs (light-activated SCRs) to drive a triac, solve the limited timer problem. A 400 VAC capacity controlling

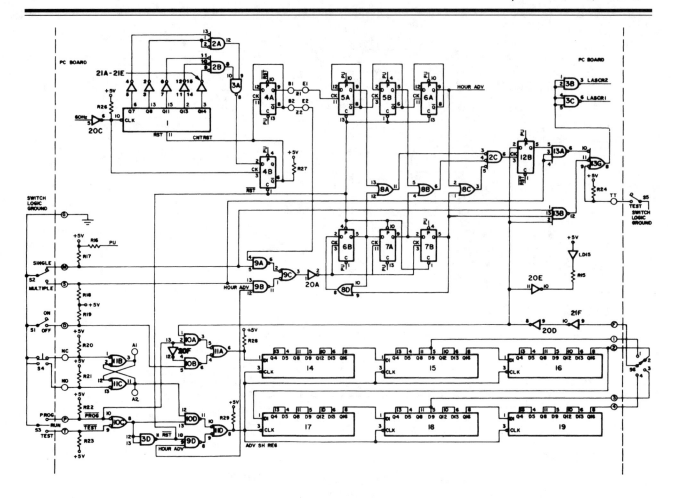

Figure 10-16. The Burglar Baffler's schematic.

your 110 VAC from the wall outlet greatly enhances dependability and reliability of operation. Refer to the Burglar Baffler's schematic and block diagram in *Figures 10-16* and *10-17*. Note the AC activation portion of the circuit.

Solving Limited Bulb Life

A zero-crossing detector protects the filament by self heating. This "electronic shock absorber" prevents applying full voltage to a cold light bulb filament. A triac applications note states a cold filament has 1/12 to 1/18 the resistance of a warm filament. If at turn-on, you are unlucky enough to catch the AC at either a negative or positive peak, approximately 170 volts appears across the filament. This is the peak (1.41) x 115 VAC. Applying 170 VAC to a resistance of only 1/18 the normal ON filament resistance results in 25 times the normally drawn current. This is a possible surge of nearly 25 amperes for a 100 Watt bulb. The protective action of the zero-crossing detector guarantees turn-on occurs at or near zero volts AC.

Solving Predictable Light Patterns

The Burglar Baffler's pseudo-random number generator controls activation times. Shift registers, acting as very simple memories, store a program of up to four days of your programmed light activations. You can make each day have a unique PNG driven pattern. The pseudo-random number generator (PNG), or linear sequence generator, is not as formidable as it sounds. The PNG is a group of serially connected flip-flops forming a shift register. The trick is to know where to tap off the shift register. These taps go to an exclusive-OR gate, and then back around to the D-input of the first

Chapter 10

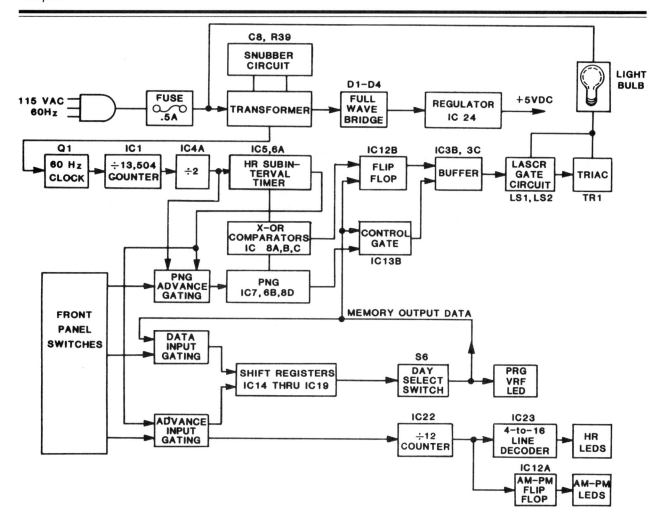

Figure 10-17. The Burglar Baffler's block diagram.

flip-flop. Correct tapping suppresses the state of all binary zeros, resulting in a totally random sequence.

Using enough flip-flops yields a truly weird sequence. We use three flip-flops, the minimum number possible, with seven distinct light activations at 1/8 to 7/8 of an hour after your selected or programmed ON hour. Using more flip-flops yields a longer PNG sequence. *Figure 10-18* illustrates this concept and where you should make the taps. After this unpredictable pattern, it recycles 28 days later. This simulates normal human behavioral patterns of staying up late at night and moving from room to room turning on lights, music, the television, etc.

The SPDT (Single-Pole Double-Throw) front panel toggle switch allows single or multiple light activations within the programmed ON hour. This adds more unpredictability because the a 3-bit PNG controls the number of Multiple Mode activations.

Operation

The front panel favors neither right-handed or left-handed people and has an orderly and easily understood logical format. There are six front panel switches. (See *Figure 10-19*) There are three toggle switches, two push-button switches, and a 4-position rotary switch. Two of the three toggle switches are SPDT (switches without a center-off position). These are labeled SINGLE/MULTIPLE and HR ON/HR OFF. They select either single or

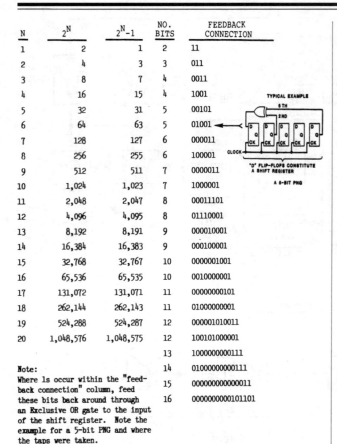

N	2^N	2^N-1	NO. BITS	FEEDBACK CONNECTION
1	2	1	2	11
2	4	3	3	011
3	8	7	4	0011
4	16	15	4	1001
5	32	31	5	00101
6	64	63	5	01001
7	128	127	6	000011
8	256	255	6	100001
9	512	511	7	0000011
10	1,024	1,023	7	1000001
11	2,048	2,047	8	00011101
12	4,096	4,095	8	01110001
13	8,192	8,191	9	000010001
14	16,384	16,383	9	000100001
15	32,768	32,767	10	0000001001
16	65,536	65,535	10	0010000001
17	131,072	131,071	11	00000000101
18	262,144	262,143	11	01000000001
19	524,288	524,287	12	000001010011
20	1,048,576	1,048,575	12	100101000001
			13	1000000000111
			14	01000000000111
			15	000000000000011
			16	0000000000101101

Note:
Where 1s occur within the "feedback connection" column, feed these bits back around through an Exclusive OR gate to the input of the shift register. Note the example for a 5-bit PNG and where the taps were taken.

Figure 10-18. Where to make PNG taps.

multiple light activations within a programmed ON hour. You use the HR ON/HR OFF switch in conjunction with the HR ADVANCE push-button to load your light activation sequence into memory.

Pressing the HR ADVANCE momentary contact switch advances to a particular hour which will be on if you have the HRS ON/HR OFF switch in the ON position. Conversely, if you select the OFF position, that hour does not activate the light bulb.

The T.T. (Triac Test) push-button tests the triac and its associated AC activation circuitry. If you plug a lamp into the Burglar Baffler and press this button, it comes on. You can't damage the bulb's filament in this test mode since the zero-crossing detector protects it. The 4-position switch called "DAYS 1, 2, 3, 4," selects how much memory you use. If you select one day, you enable 24 bits, for two days, you enable 48 bits, up to four days and

96 bits. This corresponds to one bit per hour. The last center-off toggle switch (RUN/PROG/TEST) sets the PROG position. You may program the timer to control light activations each hour of the day, even if you select multiple days.

Manual Programming

You merely repeatedly press the HR ADVANCE push-button switch while setting the HR ON/HR OFF switch to the position of light activations you desire within that hour. For example, if you are at 11 AM and you have the HR ON/HR OFF switch ON, the light comes on during the 11 AM hour. When in the PROG mode don't worry about a previously entered program because you are pushing out old data as you open the loop. New data replaces this old data. This is like a calculator's pushdown stack operation of entering data into registers. The TEST mode examines the new program but does not open the loop.

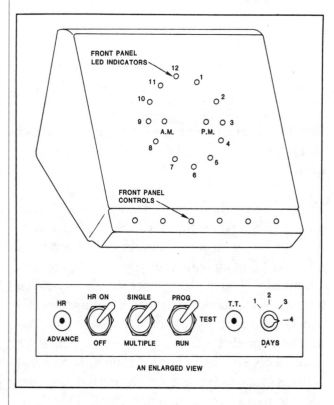

Figure 10-19. The Burglar Baffler's front panel switches.

The Burglar Baffler's 12 circularly placed LEDs represent a clock's face. There are two more AM and PM LEDs which saves us from using 24 LEDs for a whole day. After you pass 12 o'clock, the AM and PM LEDs swap states. The fifteenth LED, labeled PRG VRF, is for program verification.

Program Verification

Once programmed, this allows you to examine the memory's contents. Your hand could have slipped during programming or you could have been momentarily interrupted by a member of your family. To verify the program, switch the 3-position toggle switch to the TEST position. If you selected the four day option, you've used all 96 hours. If you program less than 96 hours, you will cause a start at an incorrect time. Once in the TEST position, step through the program by pressing the HR ADVANCE push-button 96 times. The PRG VRF (program verification) LED lights if you programmed that corresponding hour to come on. Naturally, if the PRG VRF LED remains off during your test, lights within that hour remain off. Verification is as quick as you can press the HR ADVANCE push-button switch — you'll see the LEDs zing around on the clock face. After entering and checking the program, place the switch in the RUN mode. This starts the timer, keeping track of how much time has passed.

For ease of explanation, let's enter a very simple program. This example still shows how to program the timer. Once you've mastered this basic procedure, you can then enter as complicated a light activation program as you wish. Assume it is 10 minutes before 5 PM. You'll push the HR ADVANCE push-button twice until it is at 7 PM. Now switch the HRS OFF/HRS ON switch to the HRS ON position. This means at 7 PM, or during that hour, the lights come on. Press the HR ADVANCE push-button switch twice until it is at 9 PM. That will cause the lights to remain on at 8 PM. If the HRS OFF/HRS ON switch is switched to the HRS OFF position the lights will turn off sometime during the 9 PM hour. Pushing the HR ADVANCE push-button twice again with the HRS OFF/HRS ON switch OFF causes the 9 PM and 10 PM hours to be off. Now we are at 11 PM, and we want the lights to be on during the next two hours. First make certain the HRS OFF/HRS ON switch is in the ON position. Now press the HR ADVANCE push-button twice. During the last hours of 1 AM to 7 PM you may wish the lights to remain off. Make certain the HRS OFF/HRS ON switch is in the HRS OFF position and then push the HR ADVANCE push-button 18 times. This ensures the next 18 hours remain off. You could enter the same pattern the second day, or change it.

You may ask, "Why make it such an obvious pattern by repeating it the second day? Isn't the predictability of light timers one of their drawbacks?" A good point! But don't forget the pseudo-random number generator controls the times during the selected hour the lights go on and off. It also controls and varies the number of times they come on within that programmed ON hour.

Before worrying about the pattern's predictability, which is no problem at all, let's verify the program. All you need is to set the PROG/RUN/TEST switch to the TEST position. Now press the HR ADVANCE push-button and watch the LEDs zing around the Burglar Baffler's clock-like face. When verifying a program, three LEDs come on simultaneously, indicating the electrical device plugged into the Burglar Baffler will come on. These LEDs are:

1. The particular hour LED being checked.
2. The PM or AM LED.
3. The PRG VRF LED.

Block Diagram

The power supply (See *Figures 10-16* and *10-17*) powers both TTL and CMOS logic. This requires a +5 VDC at 500 mA DC regulated source. You also need a 12 VAC source which you easily derive from the transformer's secondary. Transformers rated at 9 to 18 VAC at 1A are fine. The regula-

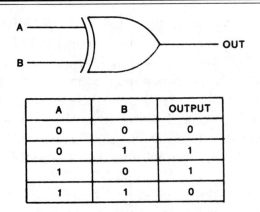

Figure 10-20. An X-OR gate's function.

tor input withstands from 7 to 34 VDC. Diodes D1 through D4 form the full wave bridge rectifier and should have a 50 V PRV minimum. You'll need a 1/2 A fuse for safety. The transformer primary has an RC network or "snubber circuit" (C8, R39) to prevent inductive coupling of AC line transients through the transformer secondary to other parts of this project. Capacitor C7 is the input smoothing or filter capacitor.

IC24 is the +5 VDC voltage regulator. Use a TO-220 plastic or ceramic 3-lead tab device. T.O. stands for "transistor outline," the term still survives in this IC era. It describes an active component's physical outline and dimensions. The regulator provides a very stable +5 VDC output, regardless of sudden changes in current. Capacitor C6 bypasses any ripple to ground that might appear at the +5 VDC output.

The 60 Hz clock circuit provides critical system timing. You derive this from the utility company's ultrastable 60 Hz, 115 VAC wall outlet frequency. The clock consists of transistor Q1, R36, C5, D9, and R25. You apply this 17 volt peak voltage from the 12 Vrms transformer secondary to another RC network (R36 and C5). This low pass filter rolls off at 60 Hz to reduce transients. Diode D9 provides noise immunity. The clipper circuit purposely overdrives Q1's base. This cuts off the top of the sinewave, forming a slow rise and fall time squarewave. Applying transistor Q1's collector to

inverter 20C improves this previously "rounded" squarewave's shape (or its rise and fall times), giving it a Schmitt trigger effect. The counter is mainly composed of IC1, a 14 stage CMOS binary counter.

The CMOS counter, (IC1), has 14 serially connected negative edge triggered "D" flip-flops with a common reset. This common reset provides a convenient means of recycling. The counter's selected outputs go high and are gated back to the reset pin to reset all outputs to zero. This enables counting through another predetermined number of pulses. This number is 13,504 which, in terms of a 60 Hz wall frequency, converts to 3 minutes and 45 seconds or 1/16 hour.

After 13,504 pulses, the counter, (IC1), has outputs Q7, 8, 11, 13, and 14 all high. These are buffered and inverted by inverters IC21A to IC21E. These low outputs from the inverters are again inverted by NOR gates IC2A and IC2B. NAND gate IC3A provides another inversion. This signal drives flip-flop IC5B and then flip-flops IC5A, IC5B and IC6A. This forms a divide-by-eight ripple through counter (the hour subinterval counter). This circuit recycles every hour, advances by one count every 1/8 hour, and applies the three flip-flop's Q outputs to exclusive-OR's IC8A, IC8B, and IC8C.

An exclusive-OR gate (See *Figure 10-20*) is actually a digital comparator. When both inputs are either simultaneously high or low, a logic low results. This is convenient, because during the hour, each time an eighth of an hour transpires, flip-flops IC5A, IC5B, and IC6A produce a different binary number from 000 to 111. This 3-bit number is compared with the 3-bit number, representing the output of the PNG (IC6A, IC7A, and IC7B). The three exclusive-OR gates have all logic low outputs when the two 3-bit numbers match. This is the only combination which forces the three input NOR gates (IC2C) high. When IC2C goes high or on the rising edge, it clocks flip-flop IC12B. This transfers the "D" input to the flip-flop's Q output. This is true during operation in the SINGLE mode.

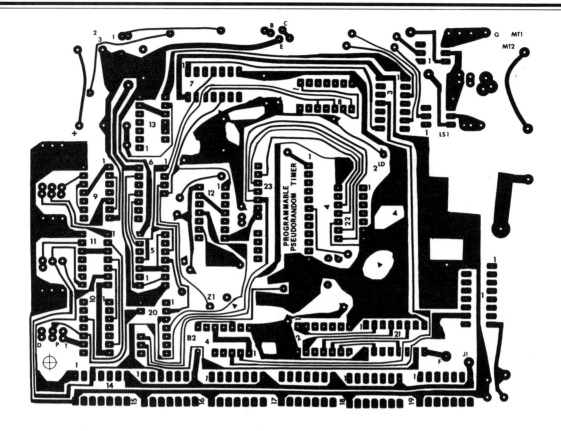

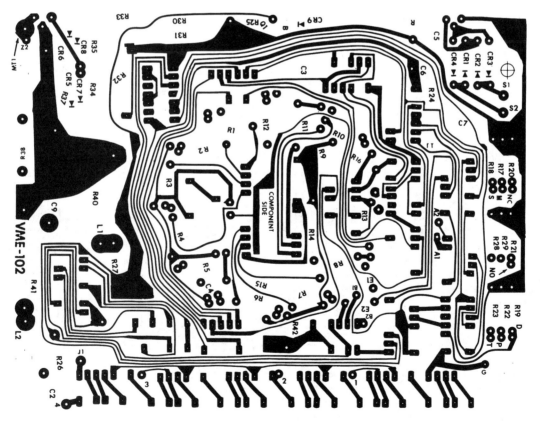

Figure 10-21. The Burglar Baffler's foil patterns.

During operation in the MULTIPLE mode, this inhibits the hour subinterval counter's output but still advances each 1/8 hour; however, no comparison is made between that 3-bit word and the PNG's 3-bit word.

In the MULTIPLE mode, the hour subinterval counter's 3-bit word is not compared and the PNG is advanced every 7 1/2 minutes (1/8 hour) from the Q output of flip-flop IC4A. While in this mode, the R17 pull-up resistor causes a logic high to be applied to one of the three inputs of NAND gate IC13B. The other two inputs are the MSBs (Most Significant Bit) of the PNG. The output of the shift registers (ICs 14 to 19) represents a capability of storing a four day or 96 hour period.

Switch F6 provides a flexible means of altering the number of days programmed. If you prefer to have an exact non-variable fixed duration, you may hardwire between points 1 through 4 and F on the PC board. *Figure 10-21* shows the PC board foil patterns. Hardwiring a number of days less than 4 could save the number of shift registers required.

Input Gating

You load this shift register memory by input gating, with highs and lows representing hours on and off respectively. Input gating is when the PROG or TEST position NAND gate, IC10C, inhibits NAND gate IC10D. This inhibits the shift registers from advancing. Programming occurs while in the PROG position.

This advancing causes cross-coupled NAND gates IC11C and IC11B to apply a pulse to NAND gate IC10D. This toggles the shift register's clocks by

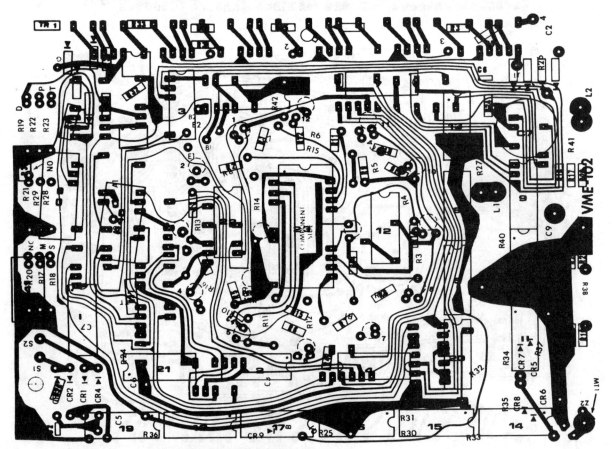

NOTE: THE 15 DOTTED CIRCLES REPRESENT LEDS WHICH ARE MOUNTED ON THE NON-COMPONENT SIDE OF THE PC BOARD. C7 MUST BE A RADIAL LEAD COMPONENT.

Figure 10-22. The Burglar Baffler's component placement drawing.

Parts List	
C1-C4	0.1 μF, 15 V ceramic or mylar
C5	0.1 μF, 25 V ceramic or mylar
C6	1.0 μF, 15 V ceramic or tantalum
C7	220 μF, 10%, 35 V electrolytic radial leads
C8, C9	0.1 μF 400 V, Sprague type 160P #4TM-P10 or Cornell Dubilier #PKM4P1
D1-D4, D9	1N4001 diode or equiv.
D5-D8	1N0004 diode or equiv.
F1	Fuse, Buss ABC5 or equiv
LED1-15	MV5054-3 LED or equiv.
LS1 (LS2)	Light Activated SCR, H11C1, H11C2 or H11C3 General Electric or MCS2 Monsanio
Q1	2N2222A
R1-R15	220W, 5%, 0.25 W
R16-25, R41	1 K, 10%0.25 W
R26-R29	4.7K, 10%0.25 W
R30, R31	270W, 5% 0.25 W
R32, R33	56K, 10% 0.25 W
R34, R35	270K, 10% 0.25 W
R36	27K, 10% 0.25 W
R37	100W, 10% 1/2 W
R38, R39	47W, 20% 1/2 W
R40	1W, 5 W wirewound
S1, S2	SPDT switch
S3	SPDT—Center off switch
S4	SPDT pushbutton switch
S5	4-position rotary switch—short between switch common and one contact at a time plus knob
T1	Transformer, primary: 117 V 60 Hz, secondary: 12.0 ± 1.0 Vac 1.0 ± 0.2 A_{RMS}
TB1	Barrier terminal block, Cinch #2-140 or equiv.
TR1	Triac, I_{av}, 8 A_{RMS}, I_{SURGE}= 100

Table 10-9. The Burglar Baffler's parts list.

NAND gate IC11D. The HRS ON/HRS OFF switch applies a logic high through gates IC10B and IC11A to the "D" input of the shift registers while in the ON position. In the OFF position, a logic low is applied to the shift register's "D" input.

The SINGLE/MULTIPLE mode switch, in the single mode, causes the AND-OR-INVERT combination of gates IC9A, IC9B, IC9C, and IC20A to make the digital comparison with exclusive-OR gates IC8A, IC8B, and IC8C. It makes no comparison in the MULTIPLE mode. In the MULTIPLE mode, the MSB of the PNG activates the light if the shift register's output through inverter IC20D is a logic high corresponding to an OFF hour. No activation occurs if one of the three inputs to gate IC13B is low.

The randomness of the hour's starting time, and the number of multiple activations, provide no set timer patterns. Also, there is a purposely induced error of approximately 2-1/2 minutes per week because of the inexact division of the CMOS counter. The counter recycles after 13,504 counts. Ideally, it should be 13,500 to be exactly 1/16 of an hour.

To more fully appreciate the randomness of the PNG counter, note the decimal equivalent of the PNG's binary output. It is 7-3-1-4-2-5-6-7-3 etc., depending upon where you enter this sequence. A PNG counter only works if you use an exclusive-OR feedback technique and suppress a state of all binary zeros. Conversely, if you use exclusive-NOR gates, you must suppress a state of all binary ones.

AC Activation

The actual turning on of an electrical device is made through a LASCR (Light Activated SCR) triac and its associated circuitry. When any of the three inputs to the 3-input NAND gate IC13C are low, the output is high. This high output is inverted by buffered power NAND gates IC2B and IC2C. They can sink twice as much current as the normal two-input TTL NAND gate. They sink 32 mA instead of the normal 16 mA. The current that they are

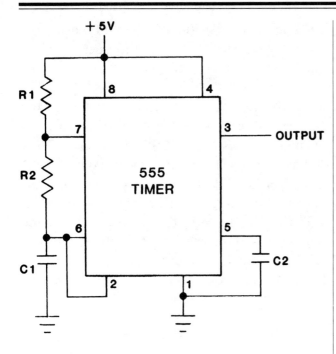

Figure 10-23. A 555 squarewave oscillator circuit.

sinking is from two LEDs. As gates IC2B and IC2C go low they cause nearly a five volt potential to develop across the LEDs of the LASCRs. These IR LEDs radiate as they conduct to trigger the IR light sensitive SCRs. Don't look for separate SCRs and LEDs. These LEDs and SCRs are housed within a 6-pin mini-DIP.

These two 200 volt LASCRs are connected in series to handle loads up to 400 volts. They are barely stressed when turning on lights, improving system reliability. Note resistors R34 and R35 are equal 270 K, 1/4W resistors. This ensures an equal division of AC voltage across the two LASCRs. The outputs of the series LASCRs, when activated, drive the triac gate. This enables it, permitting AC to flow and turns on an electrical device plugged into the Burglar Baffler.

Power Capacity

You can draw 500 Watts from the Burglar Baffler with resistive loads. Small transistorized radios and tape recorders are small enough to have their inductive (transformer) inputs effectively driven by the Burglar Baffler. If you have an old 8-track recorder it would be ideal since it uses a continuous reel which just repeats itself. Plugging a refrigerator into the Burglar Baffler would not only be ridiculous but also harmful to the Burglar Baffler since it is designed for lower-power resistive loads.

Precautions

TTL easily drives CMOS; however, CMOS will not as easily drive TTL. Driving TTL logic with CMOS requires CMOS buffers or special inverters such as CD4009, 4010, 4049, or 4050. The timer's CMOS counter uses 4009 buffered inverters (IC21A to IC21E) to drive the 7427 3-input gates (IC2A and IC2B). Another precaution when controlling a large amount of power is to attach heatsinks for the LASCR and the triac.

You should follow three rules. First, use minimum lead lengths when mounting SCRs. Second, do not connect heat dissipating devices to the SCR leads. Third, shield heat radiating devices such as lamps, power transformers, and resistors from radiating their heat directly onto the SCR case. Using the PC board foil pattern of the board avoids the majority of these pitfalls.

When consulting the parts list, the ICs that begin with 74 are TTL ICs, and the others that begin with CD40XX are CMOS ICs. Handle them with care, noting potential static electricity hazards.

Construction

Mount the transformer, assuring proper clearance, followed by the power supply circuit. Use a temporary heatsink in the form of a piece of metal, preferably aluminum or copper, attached to the hole on the +5 volt "tab" regulator. Measure the regulator's output voltage. Assemble the board's remaining components, noting the proper mounting side. (See *Figure 10-22*) The parts list is in *Table 10-9*. Do not assemble the triac or LASCR circuitry on the PC board yet. Add wires by match

Chapter 10

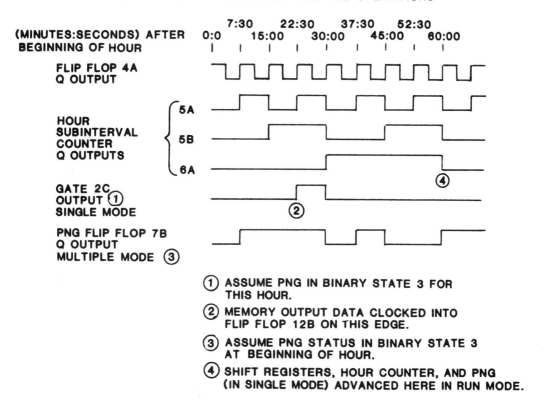

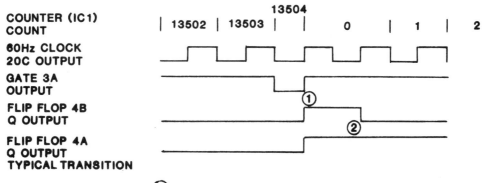

Figure 10-24. The Burglar Baffler's timing diagram.

ing B1/E1 points and then matching B2/E2 points on the PC board. Allow an adequate service loop on the switch wiring so the wires do not have to be changed when the board is mounted inside the cabinet. Perform a check on the logic. Assemble the triac AC activation circuitry. Use only a 25 or 60 Watt bulb until the triac is mounted to the cabinet. Next, mount the board to the cabinet. Mount the regulator to the chassis using a mica insulating washer, and silicone grease.

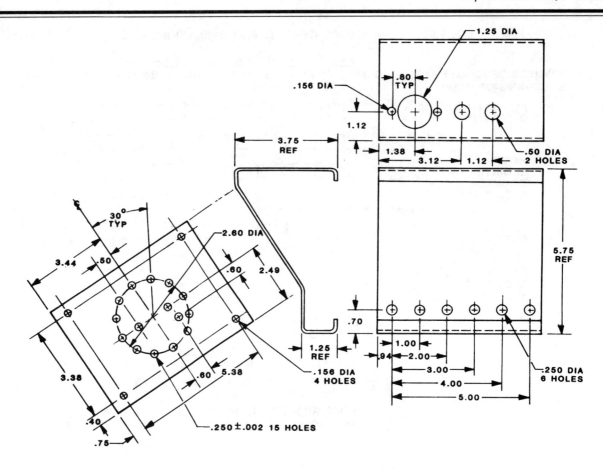

Figure 10-25. The Burglar Baffler's front panel drill drawing.

Carefully observe the triac pin configuration before mounting; thoroughly clean the surface with steel wool. This ensures flush mounting surfaces. The triac requires an insulator between it and the cabinet's surface. Use an audio oscillator and apply the output to the 60 Hz clipper input at R36. If you do not have an audio oscillator, a one-shot multivibrator oscillator can be easily constructed using a 555 timer. (See *Figure 10-23*) Apply the timer's output to inverter IC2-A's input.

Do It Yourself

If you breadboard or lay out your own PC board, observe three critical items. First, make the current carrying PC foil traces wider for the AC activation portion of the timer. Second, maintain 1/4 inch trace spacings. The divide-by-twelve IC (7492) has power on pin 5 and ground on pin 10. Third, the output of the 74154 16-to-4 line decoder/demultiplexer IC has a peculiar output. Several outputs are skipped to give a continuous count.

Initial Check Out Procedure

If you don't connect the LEDs to the correct pin, the hour activation pattern has a weird non-continuous sequence. See *Figure 10-22* again, the 15 LEDs (circular dotted-in figures) are not mounted on the component side of the PC board. They go on the non-component side and stick through the front panel. This is an exception to the standard practice. The LEDs with a number within the dotted-in circle only appear to be wrong due to their counterclockwise "mirror image."

The circular pads with a point are LED anodes, the longer of the two leads. Verify polarity on one lead to make certain it is the anode. Don't mount ICs 13 and 15 in the same orientation as other ICs.

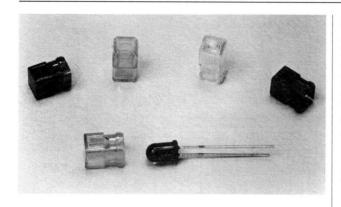

Figure 10-26. A ClipLite™ LED lens and holder. (Courtesy of Visual Communications)

Note the dot in the DIP IC's corner. The dimple next to the IC's center notch indicates pin 1.

The schematic, *Figure 10-16*, has a dotted line defining the board's edge. Circles at the board's edge with either letters or numbers within them identify wires entering or leaving the board. There are several connections and one ground in the lower right hand portion of the PC board's component side. The wiring on the board's non-component side has a pad, Z1, beside it. This is a Z-wire providing through-the-board electrical continuity. Push a straight solid strand wire through, solder both sides, and then snip it off flush.

Other on board wiring is for troubleshooting, including pad pairs (B1/E1, B2/E2, and A1/A2). If the counter and 60 Hz circuit is working, there is an output at points B1 and B2 toggling every 7-1/2 minutes (1/8 hour). This is too long, so you may wire from points A1 and A2 to points E1 and E2 respectively, producing debounced toggling outputs at points B1 and B2. Pressing the momentary contact ADVANCE switch expedites this otherwise slow process. Pulling point T.T. low tests the triac and AC activation circuits. *Figure 10-24* is the timing and troubleshooting diagram.

Tips for LED Alignment

Aligning 15 LEDs is difficult (See *Figure 10-25*) the front panel drill drawing. Try to exactly follow these dimensions to ease front panel LED alignment. Holding the LEDs is a problem until all 15 LEDs properly align. The Clip Lite™, manufactured by Visual Communications (See *Figure 10-26*), helps hold and mount these LEDs. The four holes on the corners are purposely oversized at .156" dia. This added "slop" also helps in LED alignment.

Non-Security Applications

The Burglar Baffler's non-security applications include using it with an adhesive back heater patch. The AC from the Burglar Baffler's output would heat the patch directly under the thermostat "tricking" it into not turning on the heat or turning it on less often during the night. You could also program it to go off several hours before you awake to warm the house.

Modifications

One obvious drawback to the Burglar Baffler occurs during a power failure. Unfortunately, the system goes down until the power comes back on, and when the power returns, the program that you entered in the shift registers is lost. The rudimentary memory made up of these shift registers is unfortunately a volatile memory. After power-up there is a random pattern of 1s and 0s.

You could alter the period determining circuitry to yield a longer period at one time and a shorter period at another time. Currently, we have 13,504 pulses per period instead of 13,500 (exactly 1/16 of an hour). Manually switching the outputs of IC1 into and out of inverters IC21A through IC21E would compensate for shorter winter and longer summer days. Considering how smart the Burglar Baffler already is, maybe this is more trouble than it is worth. It is only food for thought.

THE DTMF IR CAR ALARM

This microprocessor-based car alarm construction project uses IR LED sensing. It combines this with a DTMF (Dual Tone Multiple Frequency) detec-

CAR BURGLAR ALARM CONTROLLER

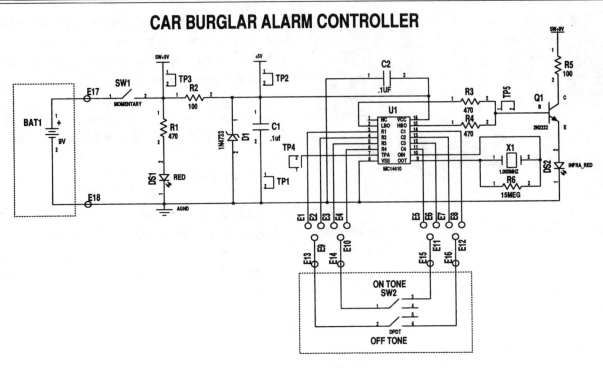

Figure 10-27. *The DTMF IR LED hand-held controller.*

tion technique. Together, it protects your car against the most common methods of attack:

1. Opening any of the car's doors, hood or trunk.
2. Motion detection of jacking up the vehicle or hit-and-runs.
3. A battery back-up option against the "pop the hood and cut the battery cable" theft.

The arm-disarm mechanism also provides fail-safe system operation.

Technical Description of the Hand-Held Controller

The hand-held controller (See *Figure 10-27*) allows you to arm (enable) and disarm (disable) the alarm system. To arm the system, point the controller at the receiver pickup and press the transmit button with the control switch in the ON position. The system answer-back LED lights for one second. This acknowledges receipt of the command. To disarm the system, point the controller at the receiver pickup and press the transmit button with the control switch in the OFF position.

After receipt of an OFF command, the answer-back LED lights for two seconds. This acknowledges receipt of the OFF command. If a door or the hood or trunk is open, the alarm will not turn on. Instead, the answer-back LED blinks nine times, awaiting further commands. After closing the door or alarm protected area, the alarm operates in a normal fashion.

The controller has a range of about three inches under normal conditions. You may achieve a longer range, depending on the level of tinting in the windshield or the ambient light at the time you hold down the transmit button. Carefully orient the controller for maximum range, the transmit and receive IR LEDs have matched narrow bandwidths.

The hand-held controller uses a 9 volt battery to power a CMOS DTMF generator. The DTMF generator's output drives an infrared LED mounted at the controller's front. A common-emitter NPN transistor buffers between the generator IC and the IR LED. The control switch is a DPDT device which selects a particular digit for the ON tone and another digit for the OFF tone. You select these to two tones for security. A visible (red) LED indicates the controller is in the transmit mode and the battery's status.

The Alarm Module

The alarm module has six circuits. (See *Figure 10-28*, the block diagram) An IR LED receiver converts the optical signal to an AC signal component on a DC voltage. The DTMF receiver decodes the AC signal into TTL compatible BCD outputs. These outputs go to the third section, a 68705 microprocessor. The signal interface circuits translate and buffer alarm inputs. The microprocessor's controlling output signals interface to relays by the relay drivers. The command code entry circuit allows selection of particular digits for alarm arming and disarming.

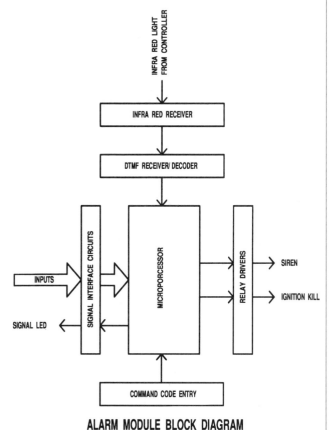

ALARM MODULE BLOCK DIAGRAM

Figure 10-28*. The DTMF IR LED alarm's block diagram.*

The Infrared Receiver

The infrared receiver is a simple two-component circuit with a 10,000 ohm resistor biasing an IR photodiode. The bias resistor's one end has +5 VDC and the other end goes to the IR photodiode's cathode. The anode of the IR LED goes to system ground. A 0.1 µF capacitor removes the signal from the IR LED's cathode. The photodiode remote mounts along with the answer-back LED. Use semirigid coaxial cable to connect the photodiode to the receiver when installing it in the car. (See *Figure 10-29*) It prevents noise, and subsequent false triggering.

The DTMF Receiver

The DTMF receiver decodes the Touch-Tone™ signals the hand-held controller generates and provides decoded outputs to the microprocessor for processing. Its input is a unity gain op amp.

The Microprocessor

The Motorola MC68705P3 Microprocessor is the alarm module controller. It monitors alarm inputs, receives decoded DTMF commands from the DTMF receiver, monitors the panic button, controls the siren and ignition cutout relays, and lights the answer-back LED, acknowledging commands. A 1 MHz crystal clocks the microprocessor. The flow chart (See *Figure 10-30*) explains the microprocessor's software operation. *Figure 10-31* is a schematic of the microprocessor circuit and the MC68705P3 microprocessor itself.

The Signal Interface Circuits

Many signals the microprocessor monitors are buffered by components within the signal interface section, which protects the microprocessor from excessive voltages (>5.0 VDC) on the input pins. These are single stage inverting PNP transistor buffers.

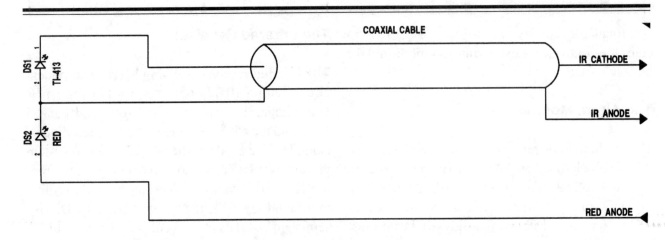

CAR BURGLAR ALARM LED WIRING

Figure 10-29. The DTMF IR LED alarm's external coaxial cable.

The Relay Drivers

The microprocessor pins assigned to control the switching relays lack drive current. Buffering them provides the necessary current. These drivers are double inverters (a PNP inverter driving an NPN inverter). This double buffering takes advantage of the B port's added current sink capability and protects the microprocessor from induced inductive switching spikes. The spike suppression diode across the relay coil also minimizes transients.

The Command Code Entry

Figures 10-32 and *10-33* are the transmitter and receiver's foil patterns, respectively. You set codes for the ON and OFF commands by inserting jumpers in appropriate places on the receiver board's component placement drawing. (See *Figure 10-34*) The inserted jumper yields a one (1); the absence of a jumper yields a zero (0). The first bank the microprocessor reads is the "ON bank," which selects the arming alarm module code. For example: 0101 = no-jumper; jumper; no-jumper; jumper = decimal 5. Consult the functional decode table, *Table 10-14*, for the DTMF decoder to ensure proper command code programming. The second bank read is the "OFF bank." You program it in the same fashion as the other bank and it selects the code to disarm the alarm module. Consult the decode table again when programming the jumpers. The ON and OFF codes must be different for proper alarm operation.

The microprocessor reads the codes during the initialization routine and stores them in memory. After an ON command, the codes stored in memory are used to determine proper command decoding.

The Construction and Operational Checkout

The initial prototype assembled used a G-10 epoxy perfboard and point-to-point wiring. After proving the concept, a printed circuit board layout followed. *Figure 10-35* is the component placement drawing of the transmitter.

Test points exist to verify proper microprocessor operation before installation in your car. Install all components, except the two ICs, on the board and apply 12 VDC to test the voltage regulator. Measure the voltage regulator's output before inserting the remaining ICs. Use temporary switches and/or short pieces of wire to simulate system hookup. If you have an oscilloscope, verify the presence of pulses at Test Point 1. Their period should be 10 milliseconds (+/- 1 millisecond). These interrupt

Chapter 10

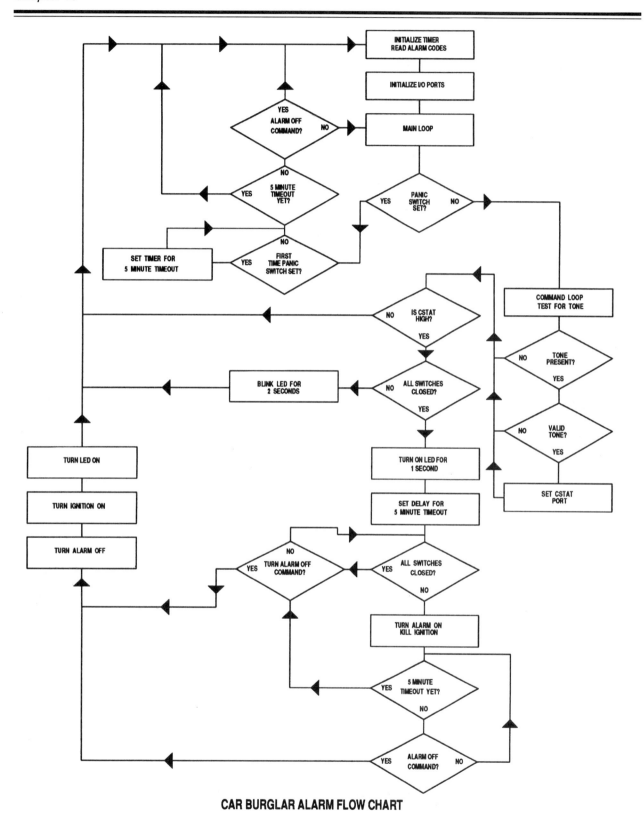

Figure 10-30. The DTMF IR LED alarm's flowchart.

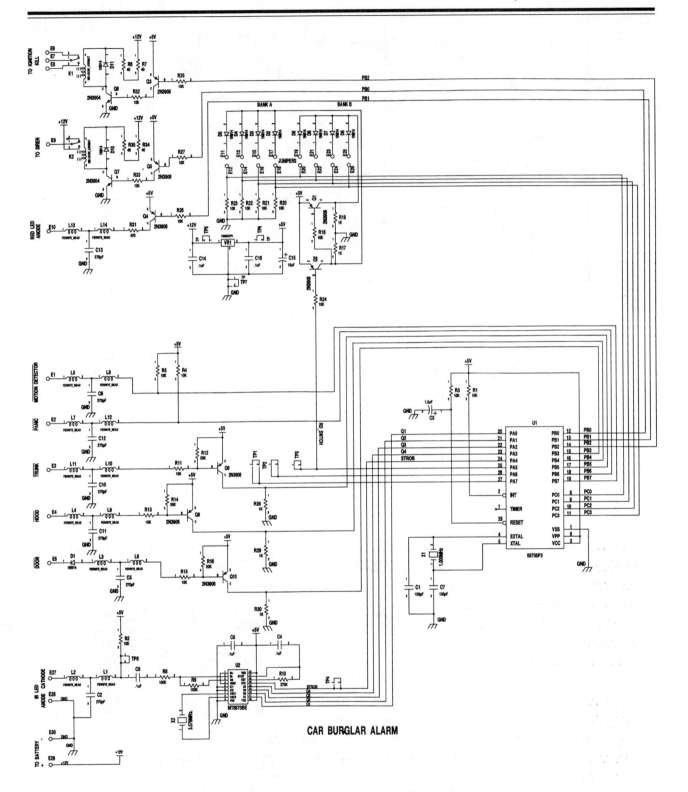

Figure 10-31. The DTMF IR LED alarm's schematic.

routine generated pulses appear on port PA7. Their absence indicates a problem with connections to the microprocessor or the IC itself.

Pulses at Test Point 2 are high when the microprocessor is executing commands within the COMMAND subroutine. Since the execution path length differs when receiving a Touch-Tone™ command,

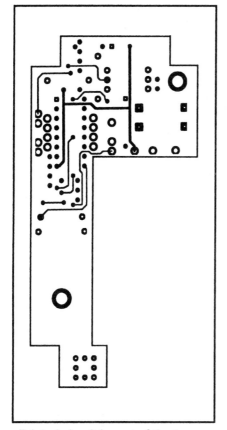 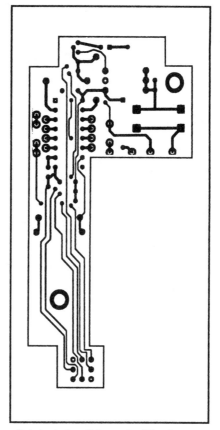

Printed circuit layout of the transmitter board (component side). Printed circuit layout of the transmitter board (solder side).

Figure 10-32. *The DTMF IR LED alarm's transmitter PC patterns.*

this yields longer pulses when the hand-held controller lights the IR receiver with IR DTMF commands. If you only have a voltmeter, monitor the voltage at Test Point 3. This point should be at +5 VDC when the microprocessor is executing instructions in the ARMED (alarm ON) loop. It is low under normal (OFF) conditions.

Send ON and OFF commands to the alarm module using the hand-held controller. Monitor the following test points for appropriate responses, making sure to correctly configure the jumpers to match the ON and OFF tones for the alarm module and hand-held controller. The alarm module does NOT respond to incorrect DTMF commands.

Test Point 1 is high (+5V) when the microprocessor is in the interrupt routine. A nominal period for this pulse train is 10 milliseconds.

Test Point 2 is high (+5V) when the microprocessor is in the command subroutine. These pulses are absent during the one and two second time delays.

Test Point 3 is high (+5V) when the alarm is ARMED and low when the alarm is OFF. The output pin of this test point also selects the command code bank the microprocessor reads.

Test Point 4 connects to the DTMF receiver's data ready strobe line, indicating the presence of valid DTMF data on the data output pins.

Test Point 5 is a + 12 VDC test point.

Test Point 6 is a + 5 VDC test point.

Test Point 7 is the system ground test point.

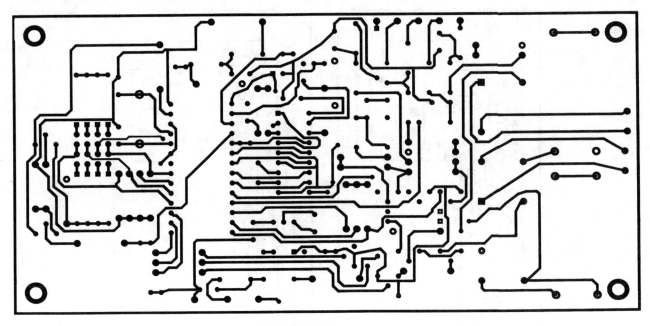

SOLDER SIDE

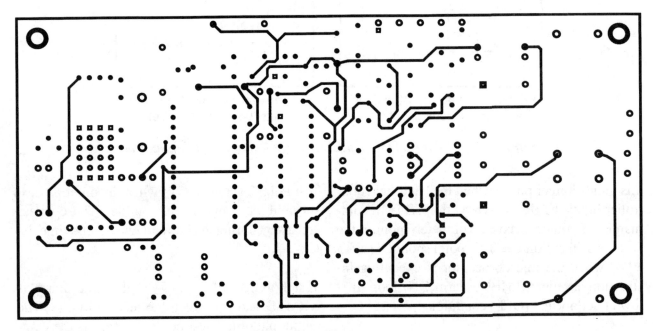

COMPONENT SIDE

Figure 10-33. *The DTMF IR LED alarm's receiver PC patterns.*

Test Point 8 tests the IR receiver's voltage output (0-5 VDC).

Installation and Operation

Mount the receiver in a location protected from excessive moisture and dust. A central location under the rear seat cushion is ideal because it requires shorter wires and is faster to install. Another good mounting location is in the trunk behind plastic or cloth/cardboard covers. Regardless, hide the receiver and all interconnecting wires to avoid thieves easily defeating the alarm. Mounting a switch on or near the motion detector per-

Chapter 10

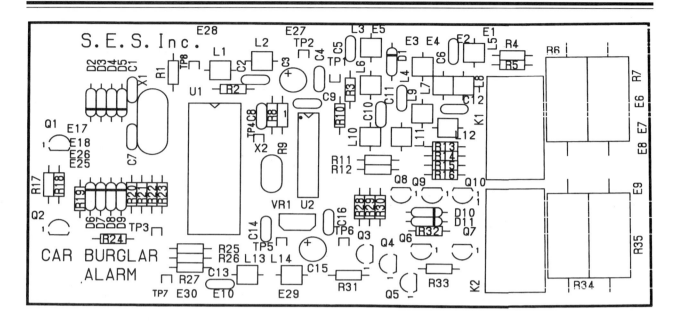

Figure 10-34. The DTMF IR alarm's receiver component placement drawing.

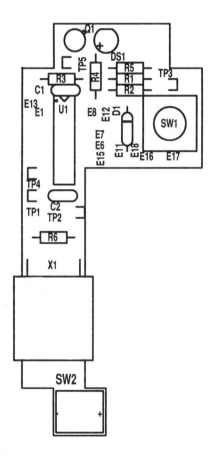

Figure 10-35. The DTMF IR alarm's transmitter component placement drawing.

mits disabling it on days windy enough days to rock your car.

Battery Back-Up Option

If you use a back-up battery, you need an isolation circuit to prevent current flow from the back-up battery flowing back into the vehicle's electrical system. (See *Figure 10-36*)

DTMF Generation and Detection

The MC14410, DTMF tone generator, (U1), produces the DTMF tones generated by the hand-held controller. The DTMF tones are composites of two separate tones. There is a low tone and a high tone. The numbers consist of tone pairs composed of one of four possible low tones and four possible high tones. (See *Table 10-10*) The MC14410 has four row inputs, R1 through R4 (pins 3 - 6), and four column inputs, C1 through C4 (pins 11 - 14). Each row selects one of the low tones and each column selects one of the high tones. By connecting one row and one column together, the IC generates 16 DTMF numbers. The MC14410 DTMF tone output chart, *Table 10-11*, shows the row and

211

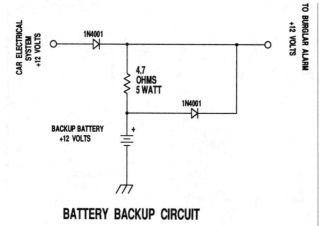

Figure 10-36. The DTMF IR LED alarm's battery back-up option.

column connections to generate the tones for each of 16 different numbers.

The jumpers in the hand-held controller select rows and columns for each of two tones. The small size of this box prevents using a DTMF 12-key phone keypad, so that's why we use jumpers instead. The hand-held IR transmitter has an aspect ratio (ratio of dimensions) of a cigarette pack, but has just 30% its volume. One number, selected by connecting jumpers E9 and E11, arms the burglar alarm. The other number, selected by connecting jumpers E10 and E12, disarms the burglar alarm. You may select any of the 16 numbers for the arm or disarm functions. You cannot use the same number for both the arm and disarm functions. The car alarm program only accepts different numbers for each function. *Figure 10-37* is a close-up view of this particularly dense or "busy" PC board's E11 to E24 "E" numbers. This helps in assembly and check out.

Selecting 3 to arm the alarm and 9 to disarm the alarm, requires first connecting jumper E9 to E1, and jumper E11 to E6. This selects 3 to arm the alarm. Connecting jumper E10 to E3, and E12 to E6, selects the number 9 for the disarm function.

Switch SW2 controls the selected numbers which modulate the IR LED with that particular DTMF frequency. This sends the arm or disarm number tones to the car alarm. The IR LED detector receives the light from the IR LED (transmitter). Its output drives the MT8870B DTMF tone decoder. When the tone decoder detects one valid low tone and one valid high tone, it produces a binary number in the form of high or low voltages on pins 11-14 (Q1-Q4).

The MT8870B tone decoder chart (See *Table 10-12*) shows the high and low voltage patterns generated for each of the 16 numbers it can decode. Generating the binary number pattern on pins 11-14 also generates a strobe pulse on pin 15. This indicates detection of a valid tone pair. The strobe is high as long as the input detects a valid tone pair. If more than one low or high tone is detected, this invalid tone pair doesn't generate an output strobe.

The jumpers you select to arm and disarm the alarm are jumpers installed in a binary sequence at E11 -

MC14410 DTMF TONE OUTPUTS

DTMF NUMBER	LOW TONE HERTZ	HIGH TONE HERTZ	ROW	COLUMN
1	697	1209	R1	C1
2	697	1336	R1	C2
3	697	1477	R1	C3
4	770	1209	R2	C1
5	770	1336	R2	C2
6	770	1477	R2	C3
7	852	1209	R3	C1
8	852	1336	R3	C2
9	852	1477	R3	C3
0	941	1209	R4	C1
*	941	1336	R4	C2
#	941	1477	R4	C3
A	697	1633	R1	C4
B	770	1633	R2	C4
C	852	1633	R3	C4
D	941	1633	R4	C4

Table 10-10. Frequencies derived from tone pairs.

Chapter 10

BURGLAR ALARM CONTROLLER JUMPERS

DTMF NUMBER	ARM TONES		DISARM TONES	
	JUMPER E9 TO	JUMPER E11 TO	JUMPER E10 TO	JUMPER E12 TO
1	E1	E8	E1	E8
2	E1	E7	E1	E7
3	E1	E6	E1	E6
4	E2	E8	E2	E8
5	E2	E7	E2	E7
6	E2	E6	E2	E6
7	E3	E8	E3	E8
8	E3	E7	E3	E7
9	E3	E6	E3	E6
0	E4	E8	E4	E8
*	E4	E7	E4	E7
#	E4	E6	E4	E6
A	E1	E5	E1	E5
B	E2	E5	E2	E5
C	E3	E5	E3	E5
D	E4	E5	E4	E5

Table 10-11. The DTMF tone output table.

E26. Jumpers E11 - E18 select the disarm number and jumpers E19 - E26 select the arm number. Input port C (PC0 - PC3) detects an installed jumper as a high. Without an installed jumper, input port C detects a low level. For example, to have 3 arm and 9 disarm the alarm, connect E23 to E24 and E25 to E26. Connecting E11 to E12 and E16 to E18 makes 9 disarm the alarm.

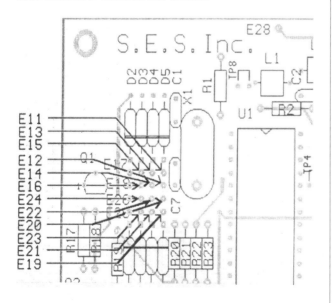

Figure 10-37. An expanded view of the PC board's E numbers.

Assembling the Transmitter's Case

An ordinary small plastic case, listed in the parts, houses the hand-held transmitter. To allow the IR LED to transmit through the case's end and have the ON/OFF switch accessible, requires modifications. See *Figure 10-38* to modify the bottom of the case. The top requires only minor modifications. The slide ON/OFF switch coming out the end opposite the IR LED has a visible light red LED built in the switch to act as:

1. A switch.
2. A power ON status indicator.

Additional Helpful Construction Hints

All L reference designators are inductors at the board's edge, where input signals enter. These ferrite beads (See *Figure 10-39*) combine with 270 pF capacitors form low pass filters (See *Figure 10-40*), eliminate hash or unwanted high frequencies. Modern cars have lots of motors to drive windows, door locks, seats, station seeking radios, antennas, mirrors, fans, and blowers, etc. These are all potential sources of unwanted "hash."

MT8870B TONE DECODER OUTPUT

DTMF NUMBER	PIN 14 Q4	PIN 13 Q3	PIN 12 Q2	PIN 11 Q1
1	LOW	LOW	LOW	HIGH
2	LOW	LOW	HIGH	LOW
3	LOW	LOW	HIGH	HIGH
4	LOW	HIGH	LOW	LOW
5	LOW	HIGH	LOW	HIGH
6	LOW	HIGH	HIGH	LOW
7	LOW	HIGH	HIGH	HIGH
8	HIGH	LOW	LOW	LOW
9	HIGH	LOW	LOW	HIGH
0	HIGH	LOW	HIGH	LOW
*	HIGH	LOW	HIGH	HIGH
#	HIGH	HIGH	LOW	LOW
A	HIGH	HIGH	LOW	HIGH
B	HIGH	HIGH	HIGH	LOW
C	HIGH	HIGH	HIGH	HIGH
D	LOW	LOW	LOW	LOW

Table 10-12. Logic levels versus decoder numbers.

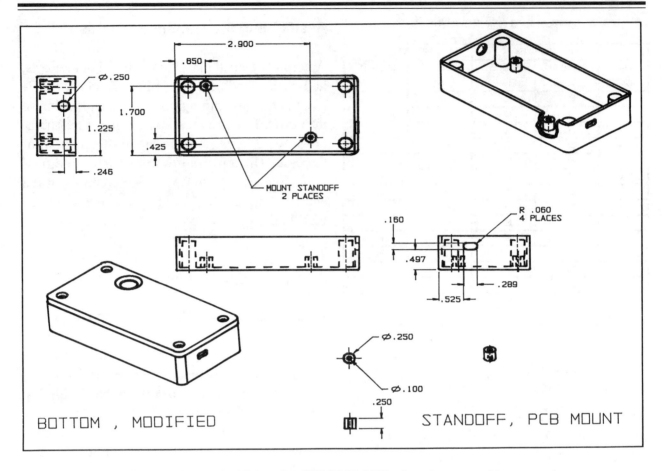

Figure 10-38. Modifying the DTMF IR LED alarm's transmitter case.

The transmitter case's lid requires a 0.250" hole. Countersink this hole with an 82° tool to slope the edge for the switch post's cap. (See *Figure 10-38* again) *Figure 10-41* shows the transmitter's drill drawing, note there are two large .140" mounting holes. *Figure 10-42* shows the Z-wires you must solder.

Figure 10-39. Ferrite beads to suppress noise and voltage spikes. (Courtesy of Fair-Rite Products, Inc.)

THE BINARY-TO-OCTAL OR HEX CODE CONVERTER PROJECT

This project uses seven ICs, a voltage regulator, a transistor array pack, 8 momentary push-button switches, a miniature toggle switch, 10 LEDs and two displays. (See *Figure 10-43* and *Table 10-13*) This project allows you to enter an 8-bit binary code by pressing switches SW1 to SW8. You may then flip toggle switch SW9 into either the octal or hex position and it displays the binary number in your selected radus or number base.

The two 74276 quad JK-flip-flops, IC U5 and U6, latch these eight inputs. The three LSBs of this 8-bit binary number go directly to display driver U4's A, B, and C inputs. The five MSBs of the 8-binary number go to the diode OR gate D1 and D2 (See *Figure 10-44*) and the CD40257 data selector, U1. This IC has two sets of 4-bit inputs, A1 to A4 and B1 to B4. You can select either set to appear at the

Chapter 10

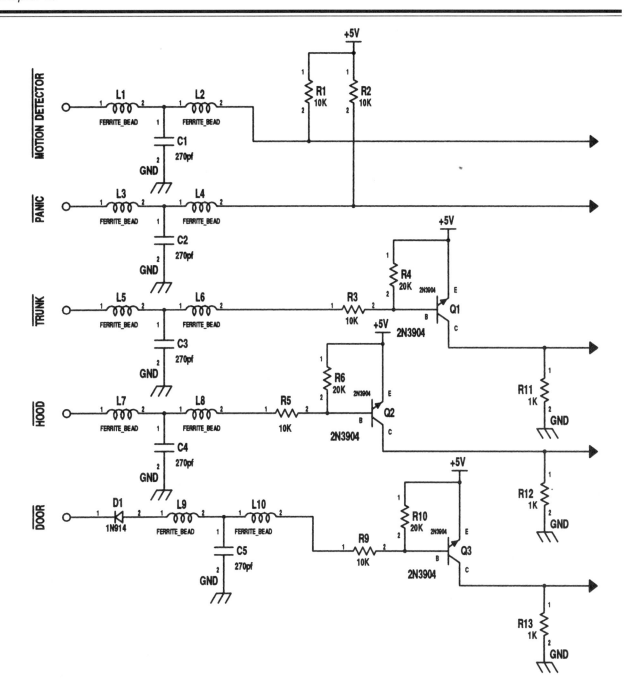

Figure 10-40. LC low pass filters.

output by controlling pin 1. If pin 1 of U1 is high, the project displays the binary number in hex and the B inputs appear as U1's output. Conversely, if pin 1 is low the A inputs of U1 appear at the output and this project displays an octal format. U8 is a quad PNP transistor array in a 14-pin DIP package with a perfect symmetry. This means it is impossible to place this IC in backwards on a PC board.

The Number Base Converter's Theory of Operation

This project displays up to 11111111 in binary which is 255 in decimal, or 377 in octal or FF in hex. Note the displays, DS3 and DS2. The lesser significant display on the right, DS3, is a dual digit display which can display up to FF. Display DS3, the more significant digit display on the left, only

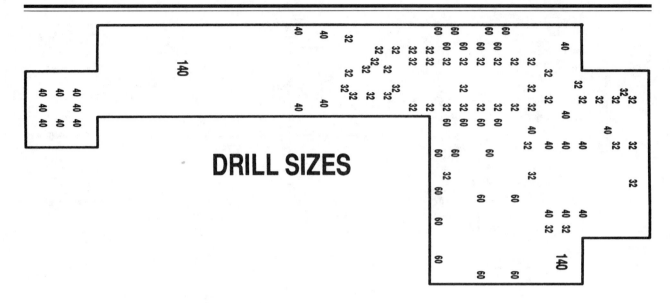

Figure 10-41. The DTMF IR LED burglar alarm's transmitter PC board drill drawing.

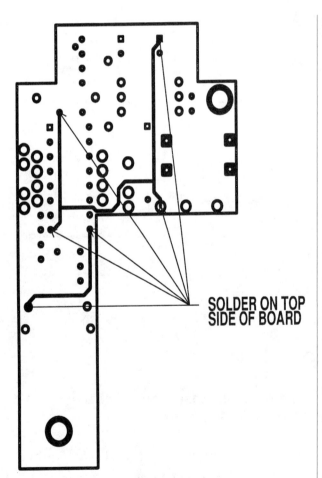

Figure 10-42. The transmitter's Z wires.

has to display up to and including 3. Therefore, IC U2 only decodes two of the possible four inputs, A and B. Inputs C and D of IC U2 are purposely tied to ground.

If you write the binary number of eight 1's and its octal and hex equivalent (See *Figure 10-45*), you'll note the different number bases just group bits into 3- and 4-bit words. Keep this in mind as we demonstrate because this is precisely what this project does through logic based grouping and routing. If you build this circuit, probe the eight data lines as you enter a binary number, first in the octal and next in hex modes. The dotted-in boxes, in line with the display drivers, show A to G. These are not actual components, just display segment designators allowing you to see which segments are active.

PC Board Making Aids

Figure 10-46 is a kit commercially available from GC Electronics to nonphotographically "lift" PC artwork patterns from these pages. You can etch a board with this film positive. Next, drill the holes for the component leads after etching. *Figure 10-47* is a process which generates PC artwork by lift

Chapter 10

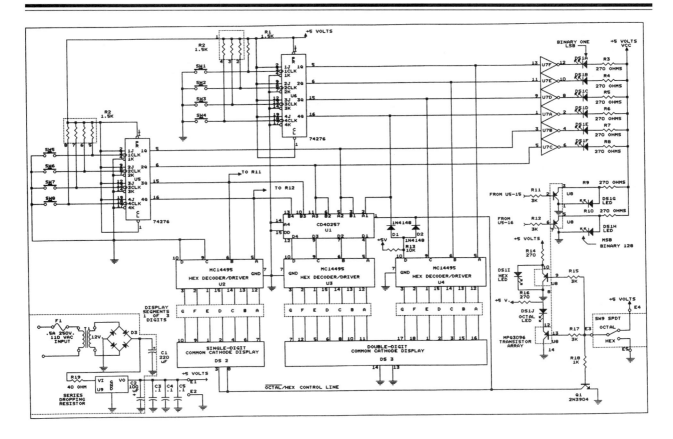

Figure 10-43. The binary-to-octal-to-hex code conversion schematic.

The Code Converter's Parts List
Resistors
R1 1.5 kW
R2 1.5 kW Seven Resistor SIP
R3 Through R10, R14, R16 270 W
R11, R12, R15, R17, 3kW
R13, 10kW
R18 1kW
R19 40W 10 Watt
Capacitors
C1 220 mF
C2 100 mF
C3 Through C5 0.1 mF
Diodes & Displays
D1, D2 1N4148
D3 Bridge Rectifier P/N LiteOn DB101
DS1A Through DS1J 10 20 mA Red LEDs
DS2 Single 7-Segment Common Cathode Display Panasonic P351
DS3 Double 7-Segment Common Cathode Display Panasonic P355
Switches
SW1 Through SW8 Momentary Contact NO Push-Button
SW9 SPDT Miniature Toggle
Fuse
F1 0.5 A 110 VAC
Transformer
12 VAC Secondary 250 mA
Transistors
U8 A Quad PNP Bipolar Transistor Array MPQ3096
Q1 2N3904 NPN Bipolar Transistor
Integrated Circuits
U1 CD40257 Data Selector
U2 Through U4 MC14495 BCD-to-7-Segment Hex Latch/Decoder/Driver
U5, U6 74276 Quad J-K Flip-Flop
U7 CD4049 Hex Inverting Buffer
U9 LM34OT-5.0 5 Volt 1 Amp Voltage Regulator

Table 10-13. The code converter's parts list.

ing foil patterns off these pages. Use LaserCoil Film by Minds in Motion, P.O. Box 697, Lindhurst, N.J. 07071. What is black on these pages becomes opaque on the finished artwork. Use your laser printer to print by centering your artwork on a semi-clear dense coating stick-backed film, such as Rayven 320 from Rayven, Inc., 431 N. Griggs St., St. Paul, MN 55104. Back this film up with a regular overhead transparency film. It won't stick to

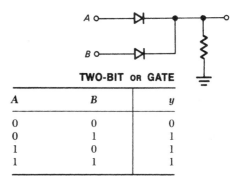

TWO-BIT OR GATE

A	B	y
0	0	0
0	1	1
1	0	1
1	1	1

Figure 10-44. A diode logic OR gate.

Take a decimal number of e.g. 117, you find its binary equivalent by successively dividing the decimal number by 2. Write a zero beside it if it is an even number and a 1 beside it if it is an odd number, completely ignoring any possible remainder. The top binary bit (beside the 117) is the LSB. The largest decimal number possible to enter is FF which is equal to 8 bits of all ones (11111111). The largest octal number is 377, the octal of FF.

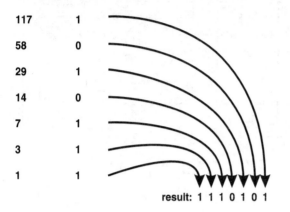

Figure 10-46. *A kit to non-photographically lift a PC foil pattern from these pages.*

Figure 10-48 is a clear Lexan™ plastic negative holding press. These two 1/4" thick layers pass light to expose photosensitive PC boards. The four corners have sets of larger and smaller holes, (4 3/8" and 8 3/4"), or a 2:1 ratio in this case. But it really depends on your negative's size. The four 0.185" diameter tools are spear pins with a paper puncher's type cutting edge. Use them on each of the larger or smaller corners. The last pieces of hardware are two 3/8" diameter pins with rolled edges. To expose negatives, remove one of these 3/8" diameter pins. This allows the top glass to pivot around the other pin's axis. Next, place your taped, aligned negative as close as possible to the press's center. Use the type of red tape that is wavelength sensitive and allows the exposure light to pass virtually unattenuated. After punching the negatives, place the unexposed board between the glass layers and punch the four corners with the cutting pins. You

To convert this number to octal just group it into three bit clusters, starting with the LSB. Hex requires grouping into four bit clusters, again, starting with the LSB.

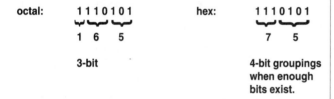

Figure 10-45. *Binary, octal and hex number equivalents.*

the printer's hot paper exposing and advancing mechanism.

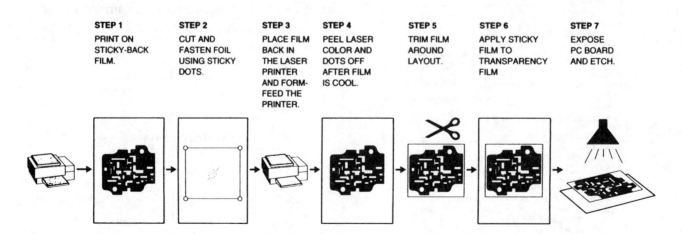

Figure 10-47. *A technique using your laser printer to non-photographically lift and make PC boards from these pages.*

can expose one side and flip it over and expose the other side of a two-sided PC board with great confidence in its precision registration.

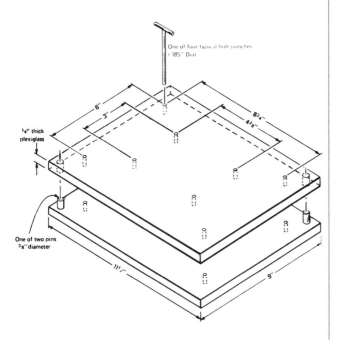

Figure 10-48. A negative holding press construction project for double-sided PC boards.

Chapter 10

Chapter 10 Quiz

1. A color temperature meter measures light intensity, T or F.
2. The human eye has color sensitive nerves (cones) for the color yellow, T or F.
3. Correcting optical filters make large shifts in color temperature, T or F.
4. Light balancing optical filters make small shifts in color temperature, T or F.
5. Which temperature scale measures color temperature?
    A. The Celsius (centigrade).
    B. The Fahrenheit.
    C. The Kelvin.
    D. None of the above.
6. The mired system is:
    A. 10^6 divided by the color temperature.
    B. The color temperature divided by 10^4.
    C. The color temperature divided by 10^5.
    D. 10^2 divided by the color temperature.
7. Which region(s) of the brain process visual data?
    A. The chromatic and achromatic regions.
    B. The achromatic region.
    C. The chromatic region.
    D. None of the above.
8. Which term defines the ability to distinguish a color from other colors?
    A. Saturation.
    B. Brightness.
    C. Hue.
    D. None of the above.
9. If you wish to present detailed visual data, which color do you not use?
    A. Green.
    B. Blue.
    C. Orange.
    D. Yellow.
10. Which portion of the eye has approximately 100 million rod-shaped and seven million cone-shaped cells?
    A. The iris.
    B. The pupil.
    C. The retina.
    D. The optic nerve.
11. Advancing the color temperature meter's momentary contact push-button switch:
    A. Regulates the battery's power.
    B. Separates light sources.
    C. Advances the dual LED's colors.
    D. "Warms" the incoming light.

12. Which color film type is balanced for heavy color temperatures occurring at noon from electronic flashes and blue flash bulbs with a 5,500 K color temperature?
    A. Tungsten.
    B. Daylight.
    C. Type A.
    D. None of the above.
13. It is more critical to obtain a true color temperature balance in color prints than slides, T or F.
14. A black body is:
    A. An electronic flash substitute.
    B. A film type.
    C. A type of camera lens.
    D. A perfect absorber and radiator of radiant energy.
15. Museums and jewelry showcases don't use ordinary fluorescent lights because:
    A. They have a light output only in certain bands.
    B. They are too bright.
    C. They tend to distort the displayed object's true color.
    D. Both A and C.
16. What purpose does the first color temperature project's 555 timer serve?
    A. It regulates positive voltage.
    B. It produces a free-running squarewave.
    C. It produces a negative voltage.
    D. It advances the dual LED's color.
17. At power turn-on, the color temperature meter initializes at one-fourth scale, indicating what?
    A. The battery is functional.
    B. The meter is working.
    C. The color temperature is at mid-range.
    D. Both A and B.
18. You should place the red end of the meter:
    A. At the lower end.
    B. At the upper end.
19. Each switch depression advances the color temperature meter how much?
    A. 1/16.
    B. 1/8.
    C. 1/4.
    D. 1/2.
20. If you elect to use a moving magnet meter it suffers from:
    A. Limited 5% accuracy.
    B. The influence of magnetic fields.
    C. Being too expensive.
    D. Both A and B.

Appendix A
An Optoelectronics Glossary

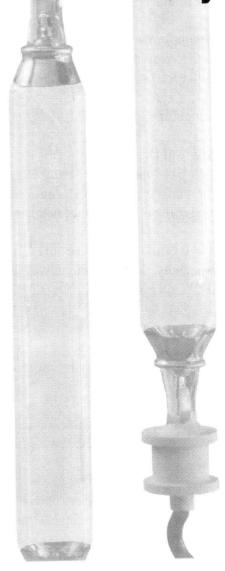

Appendix A
An Optoelectronics Glossary

Aberration — Rays from a zero-dimension object, such as a distant star, imaged through a perfect lens will all focus on a single zero-dimension spot. If the rays go anywhere else, that is an aberration.

Acceleration Factor — A factor which describes the change in a predicted phenomenon caused by a secondary effect.

Amorphous Silicon — A low cost form of silicon with a very disordered molecular structure, quite unlike the lattice molecular structure of silicon used in ICs and transistors. It is a core emerging solar technology.

Angle of Divergence — This is the spread of light after passing through a lens. The smaller the light source, the smaller the angle of divergence. A longer focal length reduces this angle.

Ångstrom — An Ångstrom (Å) measures length, especially electromagnetic wavelengths. One Å = 10^{-10} meters = 10^{-4} microns = $3.937 \cdot 10^{-9}$ inches.

ANSI — This is the American National Standards Institute, an industry-wide organization standardizing upon products, including most optoelectronic devices.

Area Source — A source with a diameter greater than 10% of the distance between it and the detector.

Avalanche Photodiode A photodiode operating on the principle of the avalanche multiplication photocurrent effect. It is very suited for applications requiring fast switching and/or low noise.

Backlighting — Illumination by an indicator of a front panel legend from behind, without the LED protruding from the front panel.

Band Gap — The potential energy difference between the conduction band and the valence band in a material. This determines the forward voltage drop and the frequency of light output of a diode.

Beam Angle — This is the angular spread of light from a lamp with angles ranging from 15° for a narrow spot to 130° for a floodlight.

Bicolor LED — A component that combines two dice of different colors upon a single substrate or lead-frame carrier. The device has either two or three lead wires for turning the device off and on.

Binary Optics — The practice of purposely cutting small grooves in the normally smooth surface of a lens. These cuts look like staircases from the side, giving a desired image in staggered segments or other unusual effects, depending on the pattern cut.

Blackbody — The device to which all irradiance measurements are referenced. A blackbody is a theoretically perfect radiator and absorber of radiant energy. Its radiation spectrum is therefore a simple function of its temperature.

Candela — A photometric unit of luminous intensity, expressed in lumens per steradian.

Candela/cm^2 — A luminance unit called a *stilb*.

Candle — This was the unit of luminous intensity for years and consisted of a wax candle about an inch in diameter. Today we better define this as one candle directed to a surface of one square foot, one foot away from the candle. It produces a uniform flow of light (flux) which is one lumen.

Candlepower — This is light intensity expressed in candles measured in one direction over a certain angle.

Chrominance Contrast — The color contrast between two adjacent surfaces of identical area, shape and texture. The human eye is more sensitive to difference in color than it is in brightness.

Collimated — The effect of concentrating a diffused light source into more of a beam shape.

Color Temperature — The temperature of a blackbody whose radiation has the same visible color as that of a given non-blackbody radiator, the measurement unit is K (Kelvin).

Concave Lens — A lens that is thinner in the middle than at its edges and tends to spread out light as it passes through the lens.

Concentrators — These emerging core solar technology devices focus sunlight on a small area (a thin strip of a silicon solar cell or a small "spot" solar cell). It converts sunlight to electricity without great quantities of silicon, lowering production costs.

Conduction Band — The empty energy band where electrons are the charge carriers.

Conservation of Energy Law — This states $E_R + E_A + E_T = 1$, or radiated plus absorbed plus transmitted energy must equal 1. An object with an emissivity of e = 0.80 has $E_A = 80\%$; therefore $E_R + E_T = 0.20$ or 20%.

Contrast — The noticeable difference in color, brightness, or other characteristics in a side-by-side comparison.

Contrast Ratio — The measurement of how visible, for example, a backlit legend is from its background. Its formula involves a ratio using both these quantities.

Convex Lens — A lens that is thicker in the middle than at its edges and tends to concentrate light as it passes through the lens.

Critical Angle — The maximum angle of incidence for which light will be transmitted from one medium to another. Light approaching the interface at angles greater than the critical angle will be reflected back into the first medium.

Crosstalk (Light Bleed) — The undesired illumination of one indicator position by an adjacent or different light source.

Crystalline Silicon — This traditional form of silicon has molecules positioned in a predetermined lattice. Crystalline structures yield higher efficiencies but cost more to produce and are soon to be abandoned in solar energy technology.

CTR (Current Transfer Ratio) — The ratio of the DC output current to the DC input current of an optically coupled isolator.

Dark Current — The leakage current of a photodetector with no radiation within its spectrum of sensitivity applied incident upon it. The smaller this leakage current, the better the optoelectronic semiconductor device.

Darlington Phototransistor — A pair of directly coupled transistors with the first transistor being a phototransistor. This device is very light sensitive.

Depletion Region — The area where the density of charge particles is negligible compared to the impurity concentration.

Detector Quantum Efficiency — The ratio expressing the number of carriers generated, divided by the number of photons absorbed.

Detector, Radiometric — A device which changes light energy (radiation) into electrical energy.

Die — The basic semiconductor device or "chip" inside the LED assembly.

Diffraction — The phenomenon of light bending at the edge of an obstacle, demonstrating wave properties of light.

Diffusant — Glass particles suspended in the epoxy lens of an LED which diffuse or broadly cast the light, increasing the viewing angle.

Diffuse Mode Photoelectric Sensors — These proximity detection photosensors are relatively inefficient, but are tolerant of the reflected beam's angle of arrival. They are very reflective sensitive, use no lens assemblies, and sense a white object at a much greater distance than a dark object, see Fixed-Field Photoelectric Sensors.

Diffusion — The clouded or scattered lens effect of an LED.

Diopter — A system which rates a lens by its refractive power instead of its focal length.

Dithering — The addition of a small signal, sometimes noise, to a system to improve performance. For example, many DVMs average a number of samples to give a more accurate measurement. Random noise helps, but a controlled signal provides better accuracy with fewer samples. This term also refers to modulation of light intensity.

Dobson Unit (DU) — A measure of how thick the ozone is in a column directly above the observer. A DU is the ozone thickness of 0.01 mm at 1°C and at 1 atmosphere. In America we experience about 300 DUs daily. It is named after the atmospheric scientist G.M.B. Dobson.

Dominant Wavelength — The wavelength that is a quantitative measurement of apparent light, as perceived by the human eye.

Doping — The addition of carrier supplying impurities to semiconductor crystals.

Dot Matrix — An array composed of addressable LED dots which can form numerous characters.

Duty Cycle — The measure of the effect of a pulsed input to a lamp, expressed as a percentage, of the ON time versus the total time.

Efficiency — The measure of the output power of a light source to its electrical input power.

Electroluminescence — The non-thermal conversion of electrical energy into light. In an LED, it is produced by electron-hole recombination in the P-N junction.

Electron Beam (EB) — This type of curing instantly cross-links adhesives, inks and coatings' polymers, more deeply penetrating a surface than mere UV curing alone.

Emissivity — The measure of how well an object absorbs or reflects radiant energy imparted upon it. Smooth shiny surfaces have emissivities approaching 0, which is a perfect heat reflector, while other objects have an emissivity value approaching 1.00, a perfect radiant energy absorber.

Emittance, Radiometric — Power radiated per unit area from a surface.

Epitaxial — Material added to a crystalline structure which has and maintains the original crystals' structure.

Erythema Action Curve — A plot, used in determining UV light and its subsequent ozone depletion's effect, of the light spectra of many biological actions.

Experimental UV Index — This Index, offered free by the National Weather Service and the EPA, is part of the weather report. It forecasts the amount of UV reaching the earth's surface in your locale at its peak hour, noon. In summer, it ranges from 0 to 15.

f Stop — This is the ratio of a lens' aperture (opening) to its focal length.

Fiber-Optics — These are optical fibers of transparent strands of glass or plastic used to conduct light energy into or out of hard to access areas. They also serve as a communications medium. A cladding material surrounds this glass or plastic, which is less dense than the cladding material and therefore has a lower index of refraction.

Fixed-Field Photoelectric Sensors — These photoelectric proximity sensors use a lens or lens assembly to purposefully limit their sensing range and compare the amount of reflected light seen by two different optodetectors. The angle of arrival of reflected light is very critical with this type photoelectric sensor.

Fixed Focus — This type photoelectric sensor detects an object in an area where emitting and receiving light cross.

Footcandles (fc) — An older irradiance (flux/area) term. A Lambert/meter^2, a more modern term, equals $9.29 \cdot 10^{-2}$ fc.

Foot-Lambert — A unit of luminance or brightness defined as the surface of one square foot upon which there is a uniformly distributed flux of one lumen, or lumen/ft^2.

Flashlamp — A device composed of glass bulbs and/or a metal can. It contains an anode, a cathode and trigger probes to guide the arc (flash). You may select the device's glass envelope and window to extend the light output into the ultraviolet range.

Flux — Power passing through a surface (energy per unit time). It is also the number of photons passing through a surface per unit time, expressed in lumens or watts.

Flux Density — The measure of a wave's strength. It is flux per unit area normal to the direction of flow, or the number of photons passing through a surface per unit area, expressed in watts/cm^2 or lumen/ft^2.

Fresnel — Augustin Fresnel (1788-1827) was a French scientist who invented a molded to shape lens, used as a light condenser, which now bears his name.

Fresnel Loss — As light passes from one medium to another, with a different index of refraction, the light lost which reflects back at this interface is Fresnel loss.

Gate — The control terminal of an SCR, or a logic function component.

H — This is an irradiance or radiation flux density expressed in watts/cm^2.

Hole Electron Pair — A positive (hole) and a negative (electron) charge carrier, considered together as an entity.

Incandescence — Emission of light by thermal excitation resulting from the superheating of a conductor. The excitation must be sufficient to produce enough photons to make the light visible.

Index of Refraction — The ratio of the speed of light in a vacuum to the speed of light through another material, e.g. a plastic lens.

Illumination — Light level on a unit area.

Infrared Radiation — The electromagnetic wavelength region between approximately 0.75 and 100 micrometers which is longer than visible light.

Injection Laser Diode — A P-N semiconductor device which uses lasing to increase the light output and concentrate the light in a small area.

Intensity — The radiant flux emitted by a light source per unit solid angle, i.e. lumens or foot-candles per steradian.

Interrupter Module — An electronic device producing an electrical output when an object breaks the path of a beam of visible or non-visible light between a photo source and detector. The emitter/detector pair is usually housed in a module with a slot.

Inverse Square Law — The illumination on a surface varies inversely as the square of the distance. As an example, a light source three feet away is only 1/9 as bright as the source at one foot.

IR LED — An infrared LED consisting of a semiconductor P-N junction emitting light, when forward biased, in the range of 0.78 mm to 100 mm.

Irradiance — The radiant flux density incident upon a surface. The ratio of flux to area of an irradiated surface, typical units are: W/ft^2 or W/m^2, with $1\ W/ft^2 = 10.764\ W/m^2$.

Isolation Voltage — The dielectric withstanding voltage capability of an optocoupler under defined conditions and time.

Joules — Energy is measured in Joules. The rate at which work is done is energy per unit time. A work rate of one Joule per second is defined as one watt. Many optical instruments collect all the light incident upon them and divides this by the detector area in cm^2, then integrate or average this over the exposure time in seconds which equals energy per unit area ($Joules/cm^2$).

Kelvin — A temperature scale starting at absolute zero where no molecular movement occurs. Zero Kelvin (no degree sign is used) is -273.15° C and each Kelvin unit equals 1° C; therefore, 0° would be 273.15 Kelvin.

(1/p) candela/cm² — A unit of luminance called a Lambert.

Lambert's Law or the Law of Cosines — This describes the spatial relationship of a perfectly diffusing surface upon which light may be either emitting or reflecting. The light emitted or accepted by the surface decreases with the cosine of the angle from a perpendicular to the surface.

Lasing — Stimulating early recognition of carriers to emit radiation in the same direction as, and coherent with, some initial stimulating radiation. It represents amplification which preserves the direction, frequency and phase of the amplified light.

LASER (Laser) — This acronym for Light Amplification by Stimulated Emission of Radiation describes a device which self stimulates and produces a very pure red light when it experiences a sufficient electric current.

Laser Efficiency — A method of evaluating the efficiency of a laser diode. Five methods are used as follows:
1. Internal Quantum Efficiency: Photons generated per current carrier injected.
2. Emission Efficiency: Photons emitted per photon emitted.
3. External Quantum Efficiency: Photons emitted per carrier current injected. This is also called "junction efficiency."
4. Differential Efficiency: The slope of the light output versus the forward current curve above the lasing threshold. Also called "slope efficiency" or "incremental efficiency."

5. Power Efficiency: Light output power divide by total input power. Also referred to as "device efficiency," "overall efficiency," or "conversion efficiency."

Lens — A curved piece of transparent material, typically glass or plastic, which bends rays of light passing through. Or, the epoxy molded to an LED die to provide a certain desired optical characteristic.

Light Current — This is the current flowing through a photosensitive semiconductor, such as a photodiode or phototransistor when exposed to illumination or irradiance.

Light Curtain — This is light produced by a line or predetermined array of emitting and receiving photoelectric sensors which detect objects within their sensing field.

Light Emitting Diode (LED) — A semiconductor diode emitting incoherent light at its P-N junction when forward biased.

Light Pipe — An optical conduit made of molded plastic that directs light from an LED to the desired viewing location, often this is at a right angle from the LED's circuit board.

Liquid-Phase Epitaxy — A process by which epitaxial are grown on substrates while immersed in liquid gallium at high temperatures.

Lumen — The unit of luminous flux measuring the flow or quantity of light. One lumen describes the light on a surface one foot square located one foot away from a one candle source.

Luminance Contrast — The observed brightness of a light emitting element compared to the brightness of the surroundings of the device.

Luminescence — The emission of light due to any cause other than temperature (heating), which is incandescence.

Luminous Flux — The time rate of flow of light. The CIE curve relates luminous flux to radiant flux, as perceived by the eye's response curve.

Luminous Intensity — Luminous flux per unit solid angle in a given direction.

Majority Carriers — Charge carriers responsible for conduction under thermal equilibrium; electrons in N-type or holes in P-type materials.

Mark Sensors — These photoelectric light sensors use an optical lamp to detect colors by distinguishing between the sensed object and its background color.

Miller Effect — The phenomenon of decreased switching speed due to inter-electrode capacitance in an optoelectronic device. Its magnitude is this capacitance times the device's gain.

Minority Carriers — Electrons in P-type material or holes in N-type material.

Mobility — The velocity of a charge carrier per unit of applied electric field.

Modulated LEDs — LEDs can turn off and on at a far greater frequency than possible with incandescent lamps. This allows the amplifier of a phototransistor receiver to "tune" to this modulated LED frequency and amplify only light signals pulsing at that frequency.

Modulation — The transmission of information by modifying a carrier signal, usually its amplitude or frequency (AM and FM).

Monochrome — Any combination of colors of the same hue, but of different saturations and luminances.

Monochrometer — An instrument which is a source of any specific wavelength of radiation over a specific band.

Monochromatic — Of a single color or wavelength.

Nanometer — A unit of length (10^{-9} meters) used as a unit of wavelength of light. It is related to the color perceived by the eye. A nanometer is equal to 10 Ångstroms.

Narrow View Reflective — This type of photoelectric sensor has a very narrow sensing field and is used for more precise detection than possible with an ordinary diffuse reflective light sensor.

Neon Lamp — A light source that generates blue or amber light by thermally exciting a neon gas plasma with heated electrodes.

Noise Equivalent Power (NEP) — This is a means of expressing an optoelectronic device's spectral response by equating it to the light level required to obtain a S/N ratio of one.

Numerical Aperture — The sine of half the angle of light acceptance.

Opposed Mode of Photoelectric Sensing — This mode of sensing has the photo-emitter and photodetector placed directly opposite each other and an object is detected when interrupting this beam of light.

Optocoupler/Opto-isolator — An optoelectronic semiconductor device transmitting electrical signals without an electrical connection between the light source (input) and a light detector (output). The input is generally an LED. The output may assume a variety of different type devices such as a photodiode, phototransistor, photodarlington pair transistor, etc.

Peak Spectral Emission — The wavelength of highest intensity of a light source.

Photoconductor — A material with resistivity which varies with changing illumination levels.

Photo Current — The difference between light current and dark current in a photodetector.

Photodarlington — A light-sensitive transistor pair connected with very high light sensitivity to illumination and radiation.

Photodiode — A semiconductor device which conducts when forward biased and when incident light falls upon its surface. The relationship between its input radiation and output current is very linear.

Photodetector — An optoelectronic device producing an electrical signal when subjected to radiation in the visible, infrared or ultraviolet regions. Photodiodes, phototransistors, and photodarlington pairs are examples of this type device.

Photometer — An instrument which measures <u>visible</u> light brightness (intensity), usually in Lumens/cm^2, and the quantity (flux) of light.

Photometry — The measurement of visible light in quantity (flux) and brightness (intensity).

Photomultiplier Tube — A photoemissive type photosensor, encased in a vacuum tube, which emits one electron per each photon falling upon a metal photo cathode. It uses successive amplification stages, using secondary emission, to amplify otherwise minute electron current.

Phonon — In the absorption process, electrons move from the valance to the conduction band, giving up heat (phonons) and light (photons). A phonon is only about 0.05 eV, 1/20th that of a photon.

Photon — A quantity (the smallest possible unit) of radiant energy. A photon carries a quantity of energy equal to Planck's Constant times frequency.

Photothyristor — A thyristor whose switching action is controlled by light applied to the thyristor's gate.

Phototransistor — A light-sensitive transistor producing an electrical signal in proportion to the intensity of the applied light. This low level photocurrent is amplified by the current gain of the transistor (or gate if the transistor is a FET). The base, or gate if it is a FET, may or may not be brought out of the case for control purposes.

Photovoltaic — A type of photosensor which generates a voltage across a P-N junction as a result of photons falling upon this P-N junction. A solar cell is a photovoltaic sensor (generator).

Photovoltaic Cell — A photosensitive device which supplies DC electricity when illuminated by radiant energy.

Photovoltaic Effect — The generation of voltage from incident radiant energy (typically the sun) from the use of dissimilar materials, one of which is light-sensitive material.

Planck's Law — This law states you can plot radiated energy as a direct function of wavelength.

Plasma — This is merely ionized gas.

Point Source — A radiation source whose maximum dimension is less than 1/10 the distance between the light source and detector.

Polarizing Filters — These antiglare filters emit only vertically polarized light. Reflected light comes back rotated by 90° so it only accepts horizontally polarized reflected light.

Prism — A device which separates light into its spectral components. In LEDs, the prism directs light output from an LED to the viewing location.

Proximity Mode of Photoelectric Sensing — This mode senses an object directly in front of it by detecting the sensor's own energy reflected back to the optodetector. This mode establishes a light beam, rather than detecting a broken beam.

Quantum Efficiency — The ratio of the number of carriers generated to the number of photons incident upon the active region.

Quantum Theory — The concept that light is emitted in minute bundles of energy rather than doled out in a steady stream.

Radiant Flux — The time rate of flow of radiant energy.

Radiometer — An instrument which measures *non-visible* light brightness (intensity), usually in Lumens/cm^2, and the quantity (flux) of light.

Radiometry — The measurement of non-visible light.

Recombination — The combining of a hole and an electron.

Reflector Module — An electronic device containing a light source and a photodetector which detects any object which reflects light back to the detector.

Refractance — The phenomenon of light bending as it passes through one medium to another, such as air to water.

Resolution — The number of visible distinguishable units in the optoelectronic device's coordinate space.

Retroreflective Mode of Photoelectric Sensing — This type photoelectric sensor contains both the photoemitter and the photodetector in the same housing. The detectors senses a reflected beam off the surface and is used to detect an object moving past, breaking this reflected beam.

Secondary Optics — Devices used to enhance or redirect an LED's light output. Examples include lenses and light pipes.

Silicon Film — This silicon, grown in long planks, is an emerging core solar technology which uses thinner silicon layers to cut costs of manufacturing.

Source — A light source which provides radiant energy.

Spatial — The directional characteristic of light in space.

Spectral Distribution — The distribution of light by wavelength with an electromagnetic spectrum.

Spectral Output — This describes the radiant energy or light emission characteristic versus the wavelength of a device.

Spectral Sensitivity — A plot of the light detector's sensitivity versus the wavelength detected.

Specular — This describes a highly reflective smooth shiny surface.

Stefan-Boltzmann Equation — Heating mass releases detectable heat, and the Stefan-Boltzmann equation describes the amount of this energy released.

Steradian — A solid angle of a sphere encompassing a surface area equal to the square of the radius of the sphere. There are 12.56 steradians on a sphere.

Temporal — The characteristic of light with time.

Thermal Detectors — Thermocouples and thermopiles are examples and these devices make radiometric measurements through the incident radiation which heats them.

Thermopile — A very broadband, heat sensing, radiation detector.

Tint — A color added to an LED's epoxy lens to identify it when it is ON.

Total Flux — This is the flux emitted in all directions. Candlepower ratings refer to flux in one specific direction.

Trapping — The capturing of a hole or an electron in an impurity or defect.

Trigonometric Reflective — These photoelectric light sensors emit light which hits the sensed object and then measure the transmitted and reflected light's difference angle.

Triple-Cell or Triple-Junction — These devices are layered three deep to purposely trap shorter wavelengths of light which were otherwise escaping. This new emerging core technology is intent on enhancing solar energy conversion efficiency.

Tungsten — The element normally used for incandescent lamp filaments. Light standards use special calibrated tungsten bulbs.

UV — This is the sun's radiant energy and occurs in three classes or radiation bands:
1. UV-A is between 320 and 400 nm.
2. UV-B is between 280 and 320 nm.
3. UV-C is between 200 and 280 nm. The shorter the wavelength, the more destructive or biologically damaging the UV.

VUV (Vacuum UV) — This UV is man-made in a vacuum and has a lower wavelength, higher frequency, and is proportionately more energetic with a far greater potential for biological damage. It is used to scribe (etch) and clean semiconductor substrates.

Valence Band — The filled energy band from which electrons are excited into the conduction band.

Wafer — A semiconductor substrate with an epitaxial layer on it.

Appendix A

Water Clear LED — An LED die combined with a clear lens that has no tinting.

Wavelength — The velocity of a wave divided by its frequency.

Wein's Displacement Law — This equation inversely relates wavelength to an object's temperature, expressed in Kelvin. This law explain the wavelength of peak energy shifting to the shorter wavelength end of the scale.

Appendix B
Sources of Supply

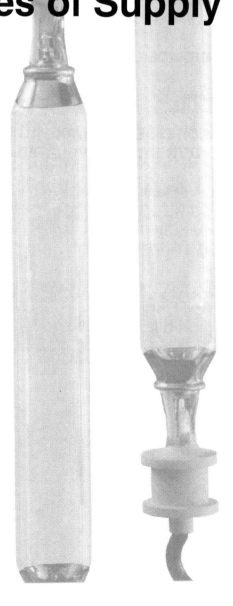

Appendix B
Sources of Supply

EXAMPLES/MANUFACTURERS OF SOLDER FLUXES/CONDITIONERS

Ultra Mild-Type R: Alpha 100
Minor-Type RMA: Alpha 611 or Kester 197
Mild-Type RA: Alpha 711-35, Alpha 809 foam, Kester 1544 or 1585
Moderate-Type AC: Alpha 830 or 842, Kester 1429/1429 foam, Lonco 3355

SURFACE CONDITIONERS

Alpha 140 or 174, Kester 5560, Lonco TL-1

ADDRESSES OF MANUFACTURERS OF LED TREATMENT RELATED PRODUCTS

Alpha Metals, Inc.
56 G Water Street
Jersey City, NJ 07304
302-434-6778

London Chemical Co. (Lonco)
240 G Foster
Bensenville, Illinois 60106
312-287-9477

The Orchard Corporation
(Silver Saver)
1154 Reco Avenue
St. Louis, Missouri 63126

Kester Solder Co.
4201 G Wrightwood Ave.
Chicago, Illinois 60639
312-235-1600

Allied Chemical Corporation
Specialty Chemicals Division
P. O. Box 1087R
Morristown, NJ 07960
201-455-5083

MANUFACTURERS OF PHOTOCELLS AND PHOTODIODES

EG&G Vactec Optoelectronics
10900 Page Blvd.
St. Louis, MO 63132
314-423-4900

Centronic Inc.
2088 Anchor Court
Newbury Park, CA 91320
805-499-5902

ADDRESSES OF IR PYROMETER MANUFACTURERS

Watlow
12001 Lackland Road
St. Louis, MO 63146
314-878-4600

Raytek
1201 Shaffer Road
P.O. Box 1820
Santa Cruz, CA 95061-1820
800-227-8074

Wahl Instruments, Inc.
5750 Hannum Avenue
Culver City, CA 90230
800-421-2853

Appendix B

Gentri Controls, Inc.
19 Ben's Way
Hopedale, MA 01747
508-634-3511

MANUFACTURERS' ADDRESSES OF IRDA RELATED PRODUCTS

ACTiSYS Corp.
1507 Fulton Place
Freemont, CA 94539
510-490-8024

Genoa Technology
5401 Tech Circle
Moorpark, CA 93021
805-531-9030

Hewlett-Packard
Corvallis, Oregon
800-677-7001

Irvine Sensors Corp.
3001 Redhill Ave.
Building III
Costa Mesa, CA 92626
714-549-8211

Puma Technologies
3375 Scott Blvd. Suite 300
Santa Clara, CA 95054
408-987-0200

TEMIC
A company of Diambler-Benz
U.S. Representative:
Siliconix
2201 Laurelwood Road
P.O. Box 54951
Santa Clara, CA 95056
408-988-8000

SOURCE OF SUPPLY FOR IR SENSOR CARDS

Laser 2000 GmbH
Argelsrieder Feld 14
82234 Wessling Germany
Tel +49 8153/405-0

Siemens Components
Optoelectronics Division
19000 Homestead Road
Cupertino, CA 95014

The UV Experimental Index

EPA Stratospheric Ozone Hotline:
1-800-296-1996

The National Weather Service:
1-301-713-0622

SOURCE OF SUPPLY FOR NIGHT GOGGLE FILTERS

WAMCO Inc.
11555-A Coley River Circle
Fountain Valey, CA 92708
714-545-5560

SOURCES OF SUPPLY FOR UV-BASED PRODUCTS

UV Xenon Flashlamps

EG&G Electro-Optics
35 Congress Street
Salem, MA 01970
508-745-3200

Hamamatsu Corp.
360 Foothill Road
P.O. Box 6910
Bridgewater, N.J. 08807-0910
1-800-524-0504

Credit Card Sized Sensometer

Mr. Rick Giese
P.O. Box 540
Kula, Hawaii 96790
SouthSky@Maui.net
1-800-96-HAWAII

UV Intensity-Light Controller

Oriel Instruments
250 Long Island Blvd.
P.O. Box 872
Stratford, CT 06497
203-377-8282
res sales@oriel.com

UV Analytical Software

Sensor Physics Inc.
attn: Gary Forrest
105 Kelleys Trail
Oldsmar, FL 34677
813-781-4240
103154.266@compuserve.com

Diesel Engine and Radiator Leak Detection by UV

UVP, Inc.
2066 W. 11th Street
Upland, CA 91786
800-452-3597

UV-Sensitive CCD Technology

PixelVision, Inc.
attn: George M. Williams
15250 NW Greenbrier Parkway
Suite 1250 Beaverton, OR 97006
503-629-3210

UV-BASED INSTRUMENTS

UV-Dosimeter

The LightBug™
International Light
17 Graf Road
Newburyport, Mass. 01950
508-465-5923

Miltec Corp.
303 Najoles Rd. Suite 108
Millersville, Md. 21108
800-999-2700

EIT Instrumentation Products
108 Carpenter Drive
Sterling, Va. 20164
703-478-0700
www.eitinc.com

UV-B Biological Effect Meter

Solar Light Company
721 Oak Lane
Philadelphia, PA 19126-3342
215-927-4206
CompuServe: 72073,2737

UV-Based Curing

Fusion UV Curing
AETEK Manufacturing, Inc.
7600 Standish Place
Rockville, Md. 20855-2798
301-251-0300

Source of UV-Based Papers

Society of Manufacturing Engineering
One SME Drive
P. O. Box 930
Dearborn, Mich. 48121
313-271-1500

Appendix B

Radtech International
North America
60 Revere Drive Suite 500
Northbrook, Ill. 60062
847-480-9080

UV-Based Flame Detectors

Spectrex, Inc.
Peckman Industrial Park
218 Little Falls Road
Cedar Grove, N.J. 07009
website: http://www.spectrex-inc.com

Detector Electronics Corp.
6901 West 110th Street
Minneapolis, MN 55438
1-800-765-FIRE

Sierra Safety Technology
702-267-2960
webmaster@flamctech.com

Spectral Sciences, Inc.
99 South Bedford Street
Burlington, MA 01803
617-273-4770
sag@spectral.com

UV Curable Adhesives

Blaze Technology Pte Ltd.
65A Jalan Tenteram #07-90
Saint Michael Ind. Estate
Singapore 328958
(65)-252-3568
email: BLAZE Technology

MANUFACTURERS OF OPTOELECTRONIC DEVICES

EG&G Sensors
Miltipas, CA 95035-7416
800-767-1888
Santa Clara: 408-988-8000
Northbrook: 847-480-9080
Cupertino: 408-257-7910
Hackensack: 201-489-8989
Palo Alto: 415-857-1501
San Diego: 619-549-6900

Scientific Technologies Inc.
31069 Genstar Road
Hayward, CA 94544-7831
800-221-7060

Siemens Components
Optoelectronics Division
19000 Homestead Road
Cupertino, CA 95014
408-257-7910

Industrial Devices, Inc.
260 Railroad Avenue
Hackensack, N.J. 07601
201-489-8989

Martech
120 Broadway
Menand, N.Y. 12204
1-800-362-9754

Texas Instruments
P. O. Box 660199
Dallas, TX 75266-0199
214-917-1264

Motorola
Literature Division
P. O. Box 20912
Phoenix, AZ 85036
1-800-441-2447
http://Design-NET.com

Hewlett-Packard
P. O. Box 10301
Palo Alto, CA 94303-0890
415-857-1501

Photocell Manufacturer

Clairex Electronics
560 South Third Street
Mount Vernon, NY 10550
914-664-6602

Optoelectronic Sensing

SUNX
1207 Maple
West Des Moines, Iowa 50265
1-800-280-6933

LED SECURING DEVICES AND LENS CAPS

Dialight
1913 Atlantic Avenue
Manasquan, NJ 08736
908-223-9400

APM Hexseal
44 Honeck Street
Engelwood, NJ 07631
201-569-5700

Bivar, Inc.
4 Thomas Street
Irvine, CA 92618
714-951-8808
bivar@interserv.com

Visual Communications Co. Inc.
7920-G Arjons Drive
San Diego, CA 92126
619-549-6900

OPTOELECTRONIC MEASURING INSTRUMENTS

Tektronix
P.O. Box 500
Beaverton, Oregon 97077
1-800-872-7924 or
1-800-547-5000

International Light
17 Graf Road
Newburyport, Mass. 01950
1-508-465-5923

PHOTODIODE AND PHOTODETECTOR MANUFACTURERS

Advanced Photonix, Inc.
1240 Avenida Acaso
Camarill, CA 93012
805-484-2884

Burle Industries, Inc.
1000 New Holland Ave.
Lancaster, PA 17601-5688
717-295-6771

Centronic, Inc.
2088 Anchor Court
Newbury Park, N.J. 91320-1601
805-499-5902

Detection Technology, Inc.
Valkjarventie 1
FIN-02130 Espoo
Finland
358 0 455 5600
info@dti.fi

Electron Tubes
100 Forge Way, Unit 5
Rockaway, N.J. 07866
201-575-5586

Appendix B

Hamamatsu Corp.
360 Foothill Road
P.O. Box 6910
Bridgewater, N.J. 08807-0910
1-800-524-0504

SOURCE FOR MOST OF THE PROJECTS' COMPONENTS

Digi-Key
701 Brooks Ave. South
P. O. Box 677
Thief River Falls, MN. 56701-0677
800-344-4539

SOURCES OF LENSES AND OPTICAL COMPONENTS

Sources of Optics Tutorials

Oriel Corp.
250 Long Beach Blvd.
P.O. Box 872
Stratford, Ct. 06497
203-377-8282

Rolyn Corp.
706 Arrowgrand Circle
Covina, CA 91722
818-915-5707

Optical Components Tutorial
Newport Corporation
1791 Deere Avenue
Irvine, CA 92606
800-222-6440
www.newport.com/tutorials/note2.html

A Convex Lenses Tutorial
Prof. Selman Hershfield
University of Florida
selman hershfield/selman@phys.ufl.edu
http://cpcug.org/user/laurence/sciteach.html

KEY SOLAR RESEARCH COMPANIES AND THEIR AREAS OF EXPERTISE

AstroPower, Inc.
Newark, Delaware
Developed silicon-film cells
Phone: 302-366-0400.

Energy Conversion Devices
Troy, Michigan
Working on roll-to-roll amorphous silicon manufacturing and triple-cell structures
Phone: 313-280-1900 or 313-362-4780.

Entech, Inc.
Dallas, Texas
Designed a new line-focus concentrator which uses 3M Linear Lensfilm to concentrate sunlight
Phone: 817-481-5588.

Siemens Solar Industries
Camarillo, California
Working on techniques to reduce the "kerf" loss or waste in sawing silicon wafers, and also in developing thin wafers
Phone: 805-482-6800.

Solarex Thin Film Division
Newtown, Pennsylvania
Developing a triple-junction thin film using less material, and realizing an enhanced efficiency
Phone: 215-860-0902

ADDRESSES OF WRIST INSTRUMENT MANUFACTURERS

Breitling U.S.A.
1-203-327-1411

Casio, Inc.
570 Mount Pleasant Ave.
Dover, N.J. 07801
201-361-5400
1-800-634-1895

Cygnus, Inc.
400 Penobscot Drive
Redwood City, CA 94063
415-369-4300
415-599-3565
FAX 415-599-2503

Polar Electro Inc.
2501 West Burbank Blvd. #301
Burbank, CA 91505
818-563-2865
FAX 818-563-2867

Seiko Communications of America
1625 NW Amber Glen Court
Ste 140
Beaverton, Oregon 97006
503-531-1623
FAX 503-531-1550

Swatch TelecomA Division of SMH (US) Inc.
35 East 21st Street
New York, New York 10010
1-800-8-SWATCH

Timex Watch Co.
Park Extension Road
Post Office Box 310
Middlebury, CT 06762
203-573-5764
FAX 203-573-4883

Appendix C
Answers to All Quizzes

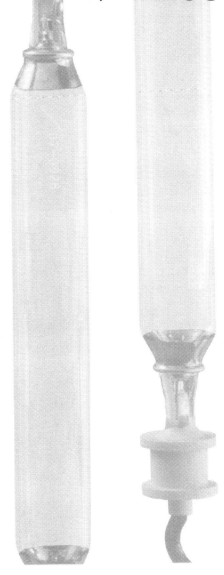

Appendix C
Answers to All Quizzes

Chapter 1 Quiz

1. T
2. F
3. F
4. T
5. T
6. F
7. T
8. T
9. C
10. D
11. A
12. D
13. D
14. B
15. D
16. D
17. C
18. A
19. A
20. C
21. D
22. A
23. T
24. F
25. T
26. D
27. C
28. A
29. C
30. B
31. B
32. D
33. B
34. A
35. D
36. C
37. D

Chapter 2 Quiz

1. B
2. D
3. A
4. D
5. A
6. C
7. B
8. B
9. A
10. T
11. F
12. F
13. F
14. T
15. T

Chapter 3 Quiz

1. C
2. A
3. A
4. D
5. B
6. A
7. C
8. B
9. D
10. C
11. D
12. T
13. T
14. A
15. F
16. B
17. C

Chapter 4 Quiz

1. D
2. A
3. A
4. D
5. D
6. B
7. C
8. B
9. D
10. D

Chapter 5 Quiz

1. F
2. T
3. T
4. T
5. C
6. D
7. B
8. A
9. T
10. D

Chapter 6 Quiz

1. B
2. C
3. B
4. A
5. C
6. C

Appendix C

7. A
8. T
9. F
10. T
11. T
12. F
13. T
14. F
15. F
16. T
17. T
18. F

Chapter 7 Quiz

1. D
2. B
3. A
4. C
5. C
6. A
7. D
8. D
9. A
10. B
11. T
12. F
13. T
14. F
15. F

Chapter 8 Quiz

1. F
2. F
3. T
4. T
5. T
6. T
7. F
8. T
9. F
10. F

Chapter 9 Quiz

1. B
2. A
3. A
4. C
5. C
6. D
7. D
8. B

Chapter 10 Quiz

1. F
2. F
3. T
4. T
5. C
6. D
7. A
8. C
9. B
10. C
11. C
12. B
13. F
14. D
15. D
16. C
17. D
18. A
19. A
20. D

Index

Index

SYMBOLS

150 DATALINK 78
150 SERIES 78
1963 CLEAN AIR ACT 9
1987 MONTREAL PROTOCOL 8
1996 IRDA STANDARD 56
3 VOLT LOGIC. 94
555 TIMER 148, 184, 185, 190
70 SERIES DATALINK 78

A

A GRAMMAR OF COLOR 176
ABSOLUTE TEMPERATURE 35
ABSOLUTE ZERO 179
ABSORPTION FILTERS 18, 19
AC CURRENT 117
AC CURVE 116
AC CYCLE 116
ACCELERATOR 14
ACCUREXTM 74
ACHROMATIC COLORS 176
ACHROMATIC REGION 175
ACID RAIN 9, 11
ACT-IR100M 66
ACT-IR200L 66
ACT-IR200M 66
ACT-IR3+ 64
ACTISYS ACT-IR3D 64
ACTISYS CORP 64
ACTISYS SYSTEMS 67
AGC 79
AGING 25
AIR CONDITIONER 12
ALARM 204
ALARM INPUTS 205
ALARM MODULE 205, 209
ALARM SYSTEM 108
ALGAE 9
ALIASING 165
ALIGNMENT 203
ALL TIME 97
ALPHA PARTICLES 26
AMBIENT INTERFERENCE 60
AMBIENT LIGHT
    61, 107, 131, 132,
    163, 164, 166, 168,
    169, 171, 174, 204

AMBIENT TEMPERATURE
    25, 119
AMPEREX 95
AMPLIFIER 58, 89, 90, 93,
    105, 140, 147, 182, 185
AMPLIFIER CIRCUIT 151
AMPLITUDE 37, 110, 134, 165,
    169
ANALOG 92, 94, 147
ANGLE OF ARRIVAL 138, 139
ANIMALS 12
ANODE 90, 109, 116, 202, 205
APERTURE 174, 189
APPLE 76
APPLE NEWTON 56, 64
APPLICATION SPECIFIC
    INTEGRATED CIRCUITS 74
ARC WELDING 15
ARF LASER 10
ARGON 179
ASIC 74, 76
ASIC IC 64
ASPECT RATIO 212
ATKINSON, JOHN S. 112
ATMOSPHERE 10, 12, 13, 16
ATMOSPHERIC PRESSURE 17
ATOMIC CLOCK 72
ATOMS 8
AUDIO EQUIPMENT 57
AUDIO OSCILLATOR 202
AUTOMATIC GAIN CONTROL 79
AUTOMATIC RANGE BUTTON
    167
AVALANCHE PHOTODIODE 105

B

BACKGROUND NOISE 43
BACKUP BATTERY 211
BACTERIA 9
BALANCED PRESSURE
    THERMOSTATIC TRAP 49
BAND GAP 105
BAND STRUCTURE 105
BANDGAPS 95
BANDWIDTH 60, 88, 93, 204
BAR AND SPACE PATTERN 70
BAR CODE 68
BAR CODE TECHNOLOGY 67
BAR GRAPH 168
BARBARELLO, JAMES J. 112
BASE 104, 107, 108, 147, 148
BASE CURRENT 107
BASE RESISTOR 104

BAUD RATE 81
BEAM 109
BEAM PARAMETERS 25
BEAM SPLITTER 26
"BELL" SHAPED CURVE 106
BELL SOUTH 75
BER 64
BER TESTER 64
BI-METAL TRAP 49
BIAS 90, 187, 205
BIMOSFET AMPLIFIER 184
BINARY CODE 214
BINARY NUMBER 212, 216
BINARY ONES 199
BINARY ZEROS 199
BIPOLAR TRANSISTOR 107
BIRDS 12
BIT ERROR RATE 64
BITS 58, 153
BLACKBODY 32, 35, 39, 179,
    180
BLACKBODY CALIBRATION 32,
    43
BLOWING AGENTS 12
BOND CLEAVAGE 8
BONDING PAD 88
BOOLEAN COMPARISON
    TECHNIQUES 16
BRAIN 175
BREITLING EMERGENCY 75
BREWER
    SPECTROPHOTOMETER 11
BRIGHTNESS 109, 166, 174,
    176, 187, 189
BRITISH ANTARCTIC SURVEY
    TEAM 12
BROMINE 12
BUFFER 111, 200
BUFFER AMPLIFIER 151
BULBS 21
BULLETIN OF SAMPLE PUBLIC
    SAFETY MESSAGES 10
BURGLAR BAFFLER
    117, 121, 191,
    192, 194, 195, 200, 203
BY-PRODUCTS 16

C

CADMIUM-SELENIDE CRYSTAL
    105
CALIBRATING IR PYROMETERS
    43
CALIBRATION 44, 77, 150, 190

Index

CAMERA OBSCURA 177
CAMERAS 23, 177
CAPACITANCE 89, 91, 96
CAPACITOR 60, 96,
    104, 117, 119, 136,
    151, 167, 184, 185,
    190, 205
CAPTURE FUNCTION 165
CAR ALARM 203, 212
CAR ALARM PROGRAM 212
CARBON BLACK 8
CARBON DIOXIDE 73
CARBON TETRACHLORIDE 12
CARS 12
CASCODE CIRCUIT 98
CASIO 83
CASIO INFRACEPTOR 82
CASIO TRIPLE SENSOR 75
CASIO WRIST REMOTE
    CONTROLLER 83
CATHODE 89, 90, 109,
    116, 205
CAVITY BLACKBODIES 44
CC FILTERS 179
CCD 23, 149
CELSIUS 33, 179
CELSIUS, ANDERS 33
CENTIGRADE 33
CFCS 12
CHARACTERS 69, 70
CHARGE COUPLED DEVICE 149
CHECK SUM CHARACTER 70
CHEMICAL REACTIONS 15
CHEMICALS 8
CHLORINE 9, 12
CHLOROFLUOROCARBONS 12
CHROMA 176
CHROMATIC COLORS 176
CHROMATIC REGION 175
CIE CURVE 140
CIRCUIT 58, 60, 88, 92, 94, 97,
    104, 108, 110, 116,
    117, 121, 122, 123, 148,
    150, 151, 160,
    167, 169, 192, 196, 205
CIRCUIT BOARD 206
CIRCUITRY 81, 96, 107, 116,
    117, 120, 165, 199, 200
CIV 91
CLEAN AIR ACT OF 1990 9
CLOSED LOOP 90
CMOS 94, 200
CMOS COUNTER 196
CMOS HEX BUFFER 120
CMOS INVERTER 111

CMR 91
COATINGS 13
COEFFICIENT OF EXPANSION
    179
COIL 88
COLLECT-BASE JUNCTION 89
COLLECTOR 89, 90, 92, 93,
    94, 96, 98, 104,
    107, 111, 148, 184
COLLECTOR CURRENT 89
COLLECTOR-BASE 89
COLLECTOR-BASE DIODE 107
COLLECTOR-BASE JUNCTION 89
COLLECTOR-TO-BASE JUNCTION
    108
COLLECTOR-TO-BASE
    RADIATION SENSITIVITY 108
COLLECTOR-TO-EMITTER
    CURRENT 108
COLLECTOR-TO-EMITTER
    SENSITIVITY 108
COLOR 107, 128, 131, 140,
    174, 175, 182
COLOR CIRCLE 176
COLOR COMPENSATING
    FILTERS 179
COLOR DISTRIBUTION 176
COLOR FILM 177
COLOR GLOBE 176
COLOR PRIMER 176
COLOR SHIFT 179
COLOR SOLID 176
COLOR TEMPERATURE
    109, 174, 177,
    178, 179, 180, 181,
    182, 189, 190
COLOR TEMPERATURE METER
    174, 182, 184, 190
COLOR TREE 176
COMBUSTIBLES 16
COMM ABORTED 77
COMM CEASED-RESEND 77
COMM DONE, DATA OK 77
COMM ERR #, SEE HELP 77
COMM ERROR-RESEND 77
COMMON MODE REJECTION 91
COMMUNICATIONS 57
COMPARATOR 90
COMPARATOR CIRCUIT 151
COMPUTERS 56
CONDUCTION ANGLE 116
CONDUCTION BAND 105, 106
CONES 174, 175
CONNECTIVITY 59
CONNECTORS 160

CONTACTS 88
CONTAINER CORPORATION OF
    AMERICA 176
CONTRAST 70
CONTROLLER 25,
    134, 153, 204, 205, 211, 212
CONVERTER 150
COOLANT 12
CORONA 91
CORONA INCEPTION VOLTAGE
    91
CORRECTING FILTERS 178
COSINE 117, 139
COTE, FLODQVIST 112
CR DIODE 21
CROSSTALK 140
CRT 76, 79, 80, 81, 93, 94
CRT DEGRADATION 95
CRYSTAL 105
CTR 91, 93, 94, 120, 121
CTR DEGRADATION 88
CURRENT 59, 88, 92, 93,
    94, 97, 98, 104,
    105, 106, 107, 108,
    109, 111, 116, 121,
    133, 147, 148,
    149, 151, 179, 185
CURRENT TRANSFER RATIO 91
CURVE 106, 108
CYCLE 8
CYGNUS 76

D

DAGUERRE, LOUIS 177
DARK CURRENT 108
DARK VOLTAGE 147
DARLINGTON CONFIGURATION
    90, 96
DARLINGTON PAIR 79, 90
DATA
    15, 26, 58, 67, 165, 170, 175, 194
DATA ENTRY 67
DATA FILE TRANSFER 62
DATA FLOW 64
DATA POINTS 169, 170
DATA RATES 81, 92, 93
DATA SHEET 91
DATA SHEETS 108
DATA SOURCES 170
DATA TRANSFERS 62
DATABASES 78
DATALINK 76, 77, 79, 80, 81

DATALINK SOFTWARE 78
DATALINK WATCHES 78
DATALOGGER 160, 171
DAYLIGHT FILM 177,
    178, 179, 181
DAYLIGHT SAVINGS TIME 79
DE-GREASERS 12
DECODER 70, 148
DECODING ALGORITHM 69
DEGRADATION 18
DELAYS 137
DENSITY 105
DEPARTMENT OF
    ENVIRONMENTAL QUALITY
    73
DESIGN TECHNOLOGY 83
DET-TRONICS 16
DETECTOR 21, 22, 88, 110,
    116, 128, 133, 160
DIATOMIC OXYGEN 11
DIE FARBENFIBEL 176
DIE FARBENKUGEL 176
DIELECTRICS 88
DIESEL PYROMETER 49
DIFFERENCE FREQUENCY
    MIXING 10
DIFFERENTIAL DISTINCTION
    SENSOR 131
DIFFUSE PATTERN 71
DIGITAL 92, 94, 146, 147
DIGITAL OUTPUT 58
DIGITAL SIGNAL 151
DIODE 58, 109, 116, 185
DIP ISOLATORS 95
DIP SWITCHES 152
DISCHARGE 91
DNA 9
DOBSON, G.M.B 11
DOBSON
    SPECTROPHOTOMETER 18
DOBSON UNITS 11
DOS 64
DOSIMETRY 25
DOW JONES 74
DRACULA 24
DREXEL UNIVERSITY 67
DRIVER 104, 119, 206
DTMF 174, 203
DTMF FREQUENCY 212
DTMF GENERATOR 204
DTMF IR CAR ALARM 203
DTMF RECEIVER 205
DTMF TONE GENERATOR 211

DUAL TONE MULTIPLE
    FREQUENCY 174, 203
DUTY CYCLE 110, 184, 186, 188
DYNAMIC RANGE 17

E

EEPROM 152, 153, 154
EG&G 71
EG&G ELECTRO-OPTICS 21
EL CHICHON 12
ELECTRICAL ENERGY 116
ELECTRICAL FIELDS 91, 107
ELECTRO-OSMOSIS 76
ELECTROMAGNETIC RADIATION
    36, 175
ELECTROMAGNETIC SPECTRUM
    8
ELECTROMECHANICAL RELAY
    88
ELECTRON 105
ELECTRON BEAM CURING 14
ELECTRONS 26, 107
EMERSON, RALPH WALDO 177
EMISSIVITY 32, 35
EMITTED ENERGY 36, 39
EMITTER 88, 90, 106, 107,
    108, 128, 129, 133,
    134, 148
EMITTER RESISTANCE 98
ENCODER 110
ENERGY 8, 20,
    58, 105, 107, 116, 179
ENERGY GAP 105
ENERGY LEVEL 8, 105
ENTRIES 78
EPA 10, 12
EPITAXIAL LAYERS 95
EPOXY 96
ERROR 43, 59
ERROR RATES 64
ERYTHEMA ACTION CURVE 18
ETI SPOTCURE 18
ETI UVICURE 18
EVALUATION BOARD
    167, 168, 170
EVM BOARD 166
EXCITATION 8
EXCLUSIVE-OR GATE 196
EXOSPHERE 10
EXPERIMENTAL UV INDEX 10
EXPOSURE TIME 21
EYE 174, 175, 176, 186

F

FAHRENHEIT 33
FAHRENHEIT, GABRIEL 33
FALKLAND ISLANDS 12
FALL TIME 96
FAST FOURIER TRANSFORM
    168
FAST IR ADAPTER 66
FEEDBACK
    92, 97, 109, 147, 185, 199
FET 60
FIBER OPTIC SENSORS 132
FIBER OPTICS 140
FILM 178, 179, 217
FILTER 26, 129,
    147, 160, 165, 166,
    168, 178, 189, 190, 213
FILTER CAP 166
FIRE 15, 16
FIRE DETECTORS 14, 16
FIRE SUPPRESSANTS 12
FIRST READ RATE 68
FISHER-PRICE 24
FIXED FIELD PROXIMITY
    SENSORS 138
FIXED-FOCUS SENSOR 131
FLAME DETECTORS 15
FLASHBULB 178
FLASHLAMPS 14
FLICKER 164, 171, 186
FLICKER FREQUENCY
    ANALYSIS 16
FLIP-FLOPS 148, 182, 192, 193
FLOAT AND THERMOSTATIC
    TRAP 48
FLORENTINE THERMOMETER
    32
FLOW CHART 205
FLUORESCENT LIGHTS 179, 180
FLUOROPTIC PROBE 26
FLUOROPTIC THERMOMETER
    25
FLUX 68, 166
FM RECEIVER 74
FOIL PATTERNS 206
FOOTCANDLE 139, 171
FORBIDDEN REGION 105
FORWARD COUPLING 91
FORWARD CURRENT 92, 120
FOSSIL FUELS 11
FOVEA 174
FOVEAL REGION 176

Index

FREQUENCY 8, 93, 104, 150, 164, 165, 168
FUEL 15
FURNITURE FINISHING 9

G

GAALAS EMITTER 95
GAALAS IR LEDS 58
GAAS IR LED COUPLER 88
GAASP EMITTER 95
GAASP LED 18
GAIN 88, 90, 93, 94, 104, 107, 108, 110, 123, 162, 167, 169, 185, 187, 190
GAIN ELEMENT 90
GALILEO 32
GAMMA RAYS 8
GAS 10
GASES 10
GATE 94, 108, 109, 116, 122, 167
GENERATOR 192
GENOA TECHNOLOGY 63, 66
GLASS 107
GLUCOPAD 76
GLUCOSE 76
GLUCOWATCH 76
GRAPH 70
GRAYBODIES 35
GROUND 88, 116, 205
GROUND LEVEL OZONE 11
GROUND LOOP CURRENTS 88, 93
GROUND LOOPING 91
GROUND LOOPS 93

H

HALF WAVE RECTIFIER 122
HALLEY BAY 12
HALO CARBONS 11
HALONS 12
HARDWARE 61, 165, 218
HARVARD UNIVERSITY 45
HAWAII 22
HCFCS 12
HEALTHY ENGLISHMAN 33
HEART RATE 169
HEART RATE MONITORS 74, 160, 168
HEAT 15, 18

HEATSINK 200
HETEROJUNCTION 95
HEX CODE CONVERTER 174
HIGH SPEED DATA SYSTEM 74
HITLER 176
HOBBIES 9
HOLE 105
HOLE-ELECTRON PAIRS 105, 107, 147
HOST 66
HP 95LX 56
HP HSDL-1000 59
HP IC 59
HSDS 74
HUE 175, 176
HUNCHBACK OF NOTRE DAME 177
HYBRID DEVICE 89
HYDROCARBON 16
HYDROCHLOROFLUOROCARBONS 12
HYDROGEN 10, 12
HYPERGLYCEMIA 76
HYSTERESIS FUNCTION 151

I

IC 89, 121, 147, 151, 166, 208
IC SUBSTRATE 88
ICON DISPLAYS 167
ICS 67, 147
IDEAL GAS 179
IGUANAS 12
IMPEDANCE 93, 104
IMPORTING DATA 78
IMPURITIES 106
INCANDESCENT 88
INCANDESCENT LAMP 117, 149
INCANDESCENT LIGHT 59
INCIDENT ANGLE 131
INCIDENT LIGHT 70, 139, 150
INCIDENT RADIATION 108, 147
INDUCTOR 117
INFORMATION ACCESS SERVICE 63
INFRARED DATA ASSOCIATION 56
INFRARED LIGHT 8
INFRARED RECEIVER 205
INFRARED SENSOR CARD 134
INKS 9, 13, 14
INPUT 96, 97, 98, 121, 165, 199
INPUT CIRCUIT 122
INPUT CURRENT 120

INPUT GATING 198
INPUT PINS 205
INPUT SIGNAL 138, 213
INPUT VOLTAGE 120, 161, 186
INRUSH CURRENTS 117
INSTRUMENT 14, 18, 19
INSULATION 88, 91
INTENSITY 18, 25, 105, 109, 132, 137, 139, 140, 166, 168, 174, 190
INTENSITY-TO-WAVELENGTH CURVE 106
INTERNAL CIRCUIT 137
INTERNAL OSCILLATOR 152
INTERPLANETARY SPACE 10
INVERTED BUCKET 48
INVERTERS 206
INVERTING MODE 153
IR 15, 19, 58, 61, 66, 88, 97, 116, 121, 139, 147
IR BASED PRODUCTS 71
IR BEAM 83
IR BEAM DETECTORS 16
IR COMMUNICATIONS 56
IR CONNECTOR 62
IR DETECTOR 40
IR ENERGY 34, 66
IR LED 56, 59, 60, 68, 116, 120, 146, 151, 171, 174, 200, 203, 204, 212, 213
IR LIGHT 13, 166
IR LINK 64
IR MODULATED TRANSMITTER 146
IR OPTICAL FILTER 61
IR PERIPHERALS 56
IR PHOTODETECTOR 88, 96
IR PHOTODIODE 60, 67, 205
IR PHOTOELECTRIC SENSOR 128
IR PHOTOEMITTER 96
IR PHOTOTRANSISTOR 153
IR PYROMETER 36, 41
IR PYROMETER PRINCIPLES 32
IR PYROMETERS 25, 36, 40, 43
IR RECEIVER 56, 59, 146, 154, 209
IR REMOTE 83
IR SCOPE 66
IR SENSOR 15, 40, 148
IR SENSOR ICS 51
IR STANDARD 56

251

IR TELEPHONES 66
IR TEMPERATURE SENSORS 50
IR TRANSMITTER
    83, 148, 154, 212
IR900SW 64
IR920SW 64
IR940SW 64
IRDA 56, 57, 58, 61,
    62, 63, 64, 66
IRDA PHYSICAL LAYER
    SPECIFICATION 59
IRDA PROTOCOL 63, 64
IRDA STANDARD 58, 63, 64, 66
IRLAP TEST SUITE 63
IRLMP TEST SUITE 63
IRRADIANCE
    19, 106, 139, 162, 179
ISOLATED GROUNDS 88
ISOLATION 91, 116

J

JERROLD 83
JOULES 20
JUMPERS 213
JUNCTION 89, 95
JUNCTION CAPACITANCE. 89

K

KELVIN 34
KELVIN TEMPERATURE SCALE 179
KEYBOARDS 57
KRYPTON 10

L

LAMBDA 116
LAMBERTIAN COSINE 19
LAMP 26
LAN 76
LAPTOP 76
LASCR 199, 200
LASER 134
LASER PHOTOELECTRIC
    SENSORS 132
LASER SENSOR 134, 136
LASER SENSOR PRECAUTIONS
    136
LASERS 10, 14, 25, 67,
    128, 132
LATCHING EFFECT 109, 116
LATENCY 165
LATITUDE 17, 88
LAW OF CONSERVATION OF
    ENERGY 39
LCD 73, 76, 166
LCD DISPLAYS 19
LDR 97
LEAKAGE CURRENT 93, 94
LEAST SIGNIFICANT BIT 80
LED 88, 97, 111, 122,
    131, 151, 182, 184, 190,
    195, 202, 203
LED ALIGNMENT 203
LED DRIVE CIRCUITS 60
LED DRIVER 151, 166
LENS 16, 17, 59, 68,
    93, 107, 131, 138, 148, 174
LIGHT 105, 107, 108, 139,
    146, 147, 170, 174, 182
LIGHT ACTIVATED SCR 199
LIGHT BALANCING FILTERS 178
LIGHT BEAM 128, 148
LIGHT BUG 18, 19, 20, 21
LIGHT CURTAIN 129
LIGHT DEPENDENT RESISTOR 97
LIGHT EMITTERS 110
LIGHT INTENSITY
    131, 147, 150, 167
LIGHT MEASUREMENT SYSTEM 165
LIGHT METER 160, 174
LIGHT METER OPERATION 168
LIGHT RADIATION 116
LIGHT SENSING HEAD 25
LIGHT SENSOR TECHNOLOGIES 131
LIGHT SOURCE 128
LIGHT SPECTRUM 185
LIGHT TIMER 195
LIGHT-TO-FREQUENCY
    CONVERTER 146, 165
LIGHT-TO-FREQUENCY IC
    150, 190
LIGHT-TO-VOLTAGE OPTICAL
    SENSORS 146
LIGHT-TO-VOLTAGE SENSORS 147
LIGHTNESS 175
LIGHTNING 15
LINE COUPLER 96
LINE VOLTAGE 25
LINEAR OPERATION 147
LINEAR TECHNOLOGY LT1319 59
LINEARITY 17, 89, 131, 167
LINKLATER, RICHARD 24
LITE-PAC 21
LIZARDS 12
LOAD 104, 117, 122, 123, 166
LOAD RESISTANCE 98
LOAD RESISTOR 104
LOGIC 94, 133, 164, 184,
    185, 195, 196, 200, 201
LOGIC GATE 92, 96
LOGIC MODE 153
LOGIC STATE 92, 111, 184
LONGITUDE 17
LOOP CURRENT 97
LORD KELVIN 34, 179
LOW-PASS FILTERS 165
LSB 80
LUX 171
LUXTRON 25

M

MAC OS 62
MAGNETIC FIELD 188
MAGNETIC FLUX 91
MANUAL PROGRAMMING 194
MARK SENSORS 131, 132
MATH CORRELATION BETWEEN
    SEVERAL SIGNALS 16
MAYER, TOBIAS 176
MEDIUM 88
MEGGA-FLASH 21
MEMORY 206
MERCURY LAMPS 14
MERCURY VAPOR 179, 180
MESOSPHERE 10
MESSAGE TRANSMITTERS/
    RECEIVERS 71
MESSAGEWATCH 72, 73, 74
METER MOVEMENTS 188
METERING CIRCUIT 185
METERS 185
METHYL BROMIDE 12
METHYL BROMIDE 12
METHYL CHLOROFORM 12
MEXICO 12
MICROCONTROLLER
    57, 149, 165
MICROORGANISMS 9

Index

MICROPROCESSOR 26, 162, 167, 205, 206, 208, 209
MICROSOFT OFFICE'S SCHEDULE+ 76
MICROSOFT WINDOWS SCHEDULE+ 78
MICROTOPS 17
MICROVOIDS 91
MILLER EFFECT 89
MINERALS 9
MIRED SYSTEM 178
MIRED SYSTEM FILTERS 179
MOBILECOMM 75
MODES 128, 153
MODULATION BANDWIDTH 91
MOLDS 9
MOLECULES 8
MOLINA, DR. MARIO 12
MOMENTUM 105
MONITORS 25, 133
MONOMERS 13, 19
MORSE CODE 75, 80
MOSFET 60
MOTORS 117
MOUNT PINATUBO 11, 12
MSB 199
MULTIMETER 191
MULTIPLE MODE 199
MUNSELL, ALBERT H. 176

N

NAND GATES 167, 182, 196, 198
NASDAQ 74
NATIONAL BUREAU OF STANDARDS 72
NATIONAL WEATHER SERVICE 10
NEGATIVES 218
NEON LAMP 88, 104
NEWTON 33
NEWTON, SIR ISAAC 176
NIST 72
NITROGEN OXIDES 11
NOISE 17, 60, 90, 136, 163
NON-IR BASED PRODUCTS 71
NON-LINEARITY 89
NON-SEMICONDUCTOR VACUUM PHOTODIODE 22
NON-VISIBLE LIGHT 139, 161
NONGRAY BODIES 36
NONVOLATILE 140

NOR GATES 196
NORMAL CONTINUOUS MODE 153
NOVALOG SIRFIR. 59
NPN BIPOLAR TRANSISTOR 60
NPN TRANSISTOR 60, 204
NRZ 80
NRZ FORMAT 81
NYQUIST FREQUENCY 165

O

OFF DELAY 137
OFF DELAY TIMER 140
OHM 60, 98
OHM'S LAW 188
OIL 16
OLIGOMERS 13
OMNIBOOK 56
ON DELAY 137
ON-RESISTANCE 60
ONE-SHOT DELAY 137
OP AM 185
OP AMP 98, 110, 111, 147, 184, 185, 188, 190, 205
OP AMP BUFFER 104
OPERATING SYSTEMS 62
OPPOSED MODE 128
OPTICAL COUPLING 116
OPTICAL DECODER 69, 70
OPTICAL EMITTER 70
OPTICAL ENERGY 116
OPTICAL FIBER 26, 88, 93
OPTICAL FILTER 22, 61, 147, 160, 182, 189
OPTICAL INTERRUPTER SWITCH 147
OPTICAL LAMP 131
OPTICAL PATTERN 71
OPTICAL SCANNING 67
OPTICAL SCANNING WAND 68
OPTICAL SENSORS 15, 146
OPTICAL TRIAC DRIVERS 115, 116
OPTICAL WAND 69
OPTICALLY COUPLED TRIAC DRIVER 121
OPTO SENSOR ARRAY 146
OPTO-COUPLING MECHANISM 120
OPTOCOUPLER 88, 89, 90, 91, 92, 93, 94, 95, 96, 97, 116, 122, 148

OPTODETECTOR 116, 128, 138
OPTOELECTRIC 139
OPTOELECTRIC SENSING MODES 131
OPTOELECTRONIC DEVICES 116, 128, 171
OPTOELECTRONIC ELEMENTS 93
OPTOELECTRONIC IC 79, 146, 174
OPTOELECTRONIC IC SENSOR 167
OPTOELECTRONIC PROJECTS 173, 174
OPTOELECTRONIC SENSORS 170
OPTOELECTRONICS 8, 96, 104
OPTOELECTRONICS DEVICE 116, 171
OPTOELECTRONICS IC 160
OPTOELECTRONICS SENSOR PAIR 169
OPTOEMITTER 128
OPTOINTERRUPTER 110, 111, 151
OPTOISOLATOR 88, 91, 92
OR GATE 196
ORBIT 8
ORIEL INSTRUMENTS 25
OSCILLATING FREQUENCY 153, 154
OSCILLATOR 151, 167, 186, 202
OSCILLATOR FREQUENCY 154
OSCILLOSCOPE 206
OSTWALD, WILHELM 176
OUTPUT 96, 97, 104, 109, 111, 121, 122, 147, 148, 151, 167
OUTPUT CIRCUIT 128
OUTPUT FREQUENCY 162, 164, 166, 167, 186
OUTPUT SIGNAL 137
OUTPUT VOLTAGE 188, 200
OUTPUTS 96, 138
OXYGEN 11
OZONE 8, 9, 11, 12, 13, 14, 17
OZONE LAYER 9, 10, 11, 12

P

P-N JUNCTION 147
PACKAGED FACTS 71

PAGERS 71
PAINT 12
PANASONIC 83
PARALLEL PORT 66
PC BOARD 9, 60, 89, 174, 182, 189, 190, 200, 202, 203, 212, 215, 218
PCMIA CARD 66
PDAS 56, 59
PEAK CURRENT 60
PEAK RATE 118
PEAKING 92
PERCEPTION 175
PERIOD 8, 80, 138, 150, 164, 165, 170, 203, 206, 209
PERIODIC TRANSMISSION MODE 153
PERIPHERAL DEVICES 59
PERIPHERAL TEST 66
PERSONAL DIGITAL ASSISTANTS 56
PESTICIDE 12
PHASE 117
PHASE LOCKED LOOP 57
PHASE SHIFT 117
PHILIPPINES 12
PHILIPS 83, 95
PHOSPHOR 18, 26
PHOTO CURRENT 147
PHOTO DIODES 168
PHOTO INITIATORS 13
PHOTO PIN DIODE 58
PHOTO-FET 88
PHOTOCELLS 105, 131
PHOTOCURRENT 89, 104, 108, 151
PHOTODARLINGTON TRANSISTOR PAIRS 93
PHOTODETECTOR 88, 89, 97, 105, 106
PHOTODIODE 19, 59, 68, 88, 89, 90, 93, 104, 105, 107, 108, 131, 147, 150, 151, 160, 190, 205
PHOTODIODE DETECTOR 137
PHOTODIODE MATRIX ARRAY 150
PHOTODIODE SENSORS 160
PHOTODIODES 22, 79, 89, 93, 104, 131, 147, 162
PHOTOELECTRIC LASER SENSORS 132

PHOTOELECTRIC PROXIMITY SENSORS 138, 139
PHOTOELECTRIC SENSING 128
PHOTOELECTRIC SENSING LASER 134
PHOTOELECTRIC SENSING MODULES 70
PHOTOELECTRIC SENSORS 128, 129, 131, 137, 138, 140
PHOTOEMITTERS 88, 106
PHOTOEXCITATION 106
PHOTOGRAPHERS 181
PHOTOGRAPHIC FILM 174
PHOTOGRAPHIC SCIENCES. 71
PHOTOGRAPHY 21
PHOTOINITIATOR 19
PHOTOMETER 17, 139, 160, 166, 167, 171
PHOTON 105
PHOTON EMITTER 94
PHOTON ENERGY 105
PHOTONS 10, 14, 58, 59, 89, 91, 95, 107
PHOTOPIC RESPONSE 166
PHOTOREFLECTORS 151
PHOTOSENSOR 128
PHOTOSENSOR ARRAYS 149
PHOTOSILICON DIODES 26
PHOTOTRANSISTOR 107
PHOTOTRANSISTOR 79, 88, 89, 90, 93, 98, 104, 105, 106, 107, 108, 109, 112, 146, 147
PHOTOTRANSISTOR CONNECTION 89
PHYSICAL MEDIA LAYER 79, 80
PIGMENTS 176
PIKE'S PEAK 75
PINS 89, 96
PIONEER 83
PIVOT JEWEL 188
PIXEL INTENSITY 150
PIXELS 80, 148, 150
PIXELVISION 23, 24
PLANCK'S LAW 32, 37
PLC 133
PLETHYSMOGRAPH 168
PLOT SCALE 170
PML 79, 80
PN JUNCTION 106
PNG 192, 199
PNP TRANSISTOR 98
PNP TRANSISTOR ARRAY 215

PNP TRANSISTOR BUFFERS 205
POCKET ORGANIZERS 71
POLAR 74
POLARITY 120
POLARIZED WAVE 129
POLARIZING FILTERS 129
POLLUTANTS 9, 11
POLYMER BASED RESINS 13
POWER FACTOR 117
POWER SPECTRUM 168, 169
POWER STROBING 90
PREAMPLIFIER 151
PREAMPLIFIER CIRCUIT 151
PRECISION OPTICS 40
PRESSURE TRANSDUCER 17
PRIMARY COLORS 176, 179
PRINT REDIRECTOR 62
PROBE 26
PROFILE 62
PROGRAM 195, 212
PROGRAMMABLE LOGIC CONTROLLER 133
PROGRAMMABLE UNIJUNCTION TRANSISTOR 109
PROJECTOR 109
PROJECTS 174, 214
PROPAGATION DELAY 91, 92, 96
PROXIMITY DETECTOR 148, 160, 169, 170
PROXIMITY MODE 128
PROXIMITY PHOTOSENSOR 138
PROXIMITY SENSORS 128, 138
PSEUDORANDOM NUMBER GENERATOR 192, 195
PUEGOT SOUND AIR POLLUTION CONTROL AGENCY 73
PULSE WIDTH MODULATION 59, 184
PULSE-MODULATION 131
PULSER 97
PULSES 109, 116
PUMA TECHNOLOGIES 62, 66
PUPIL 174
PURITY 175
PUT 109
PWM 59, 184, 186
PYROMETER 36, 40, 50
PYROMETER INSTRUMENTS 50
PYROMETER SOFTWARE 50

Q

QUADRIPLEGICS 83
QUIET ZONES 69

R

R PHOTODIODES 61
R960SW 64
RADIANT ENERGY 9, 179
RADIATED ENERGY 15
RADIATION 15, 26, 70, 106, 180, 190
RADIATION PATTERN 71, 106
RADIATION SPECTRUM 139, 179
RADICALS 8
RADIO FREQUENCY INTERFERENCE 122
RADIOMETER 139, 171
RANKINE SCALE 34
RANKINE, WJM 34
RATIO OF DIMENSIONS 212
RAYTEK 45, 50
RC TIME CONSTANT 167, 186
REAL WORLD IR PYROMETERS 40
RECEIVER
    129, 133, 134, 152, 153
RECEIVING ANGLE 140
REDUCED INSTRUCTION SET CONTROLLERS 64
REFLECTANCE FACTOR 148
REFLECTED ENERGY 43, 128
REFLECTING ANGLE 131
REFLECTION 131
REFLECTIVE OBJECT SENSOR 147
REFLECTIVITY 68, 70
REFLECTOR 130
REFLEX MODE 128
REGULATOR 166
RELAYS 133
RESET TIME 167
RESISTORS 90, 96, 110, 117, 120, 147, 167, 182, 184, 185, 186, 188
RESOLUTION
    17, 69, 149, 165, 176
RESPONSE CURVE
    19, 106, 107
RESPONSE SPEED 92

RESPONSE TIME 43, 131, 132, 137, 140
RESPONSIVITY 89
RETINA 174, 175
RETROFLECTIVE MODE 128
RETROFLECTIVE SENSING MODE 128
RETROREFLECTIVE SENSOR 128, 129
REVERSE COUPLING 91, 93
RF FIELD 26
RF OSCILLATOR 57
RF PRODUCTS 71
RFI 122
RING SIGNA 97
RISC 64
RISE TIME 96, 97
ROBERT HOOK SEALED THERMOMETER 32
ROBERTSON-BERGER METER 18
ROCKS 9
RODENTS 12
RODS 174
ROM 26
ROM LOOK UP TABLE 16
ROWLAND, DR. SHERWOOD F. 12
RUNGE, PHILIP 176
RZ SIGNALS 80, 81

S

S1 FUNCTION 168
S2 FUNCTION 168
S3 FUNCTION 168
SAMPLE RATE 165, 169, 170
SATURATION 175, 176
SATURATION COLORS 176
SAW 57
SCALE FACTOR 168
SCAN RATE 18
SCAN RATES 81
SCHEMATIC 182, 203, 205
SCHMITT TRIGGER ACTION 98
SCHMITT TRIGGER EFFECT 196
SCHOTTKY TTL LOGIC 97
SCIENTIFIC ATLANTA 83
SCR 108, 109, 116, 122, 200
SCR OPTOCOUPLERS 88
SCR TRIGGERING 21
SEIKO MESSAGEWATCH 72, 73
SELF DIAGNOSTICS 136

SEMICONDUCTOR LASERS. 132
SEMICONDUCTOR SUBSTRATE 10
SEMICONDUCTORS
    96, 105, 116
SENSITIVITY
    105, 106, 107, 138, 168, 186
SENSOR HOUSING 40
SENSOR PHYSICS 25
SENSOR PLACEMENT 42
SENSOR/TARGET DISTANCE 41
SENSOR/TARGET DISTANCE RELATIONSHIP 32
SENSORS
    17, 131, 136, 137, 138, 140, 146, 148, 160, 162, 170, 191
SERIAL IR 56
SERIAL PORT 66, 67
SHARP PT370 79
SHARP WIZARD 56
SHARP WIZARD/ZAURUS 64
SHEEP 12
SIGNAL GENERATOR 151
SIGNALS 60, 88, 98, 146, 147, 165, 205
SILICA 25
SILICON 79, 147
SILICON CONTROLLER RECTIFIER 108, 116
SILICON DETECTOR PIN DIODES 58
SILICON PHOTODETECTORS 88, 95
SILICON PHOTODIODE 25, 146
SILICONE 95
SILK 14
SILK SCREEN 14
SILVER, BERNARD 67
SINE WAVE 169, 196
SINGLE-POLE DOUBLE-THROW 193
SIR 56, 60, 62
SIR TESTER 66
SLACKER 24
SLEEP MODE 164
SMOG 11
SNUBBER
    117, 118, 119, 121, 196
SNUBBER CIRCUIT 117
S0 FUNCTION 168

Index

255

SOFTWARE 25, 50,
    61, 62, 63, 64,
    66, 77, 160, 161,
    164, 165, 166, 167,
    168, 169, 170
SOIC 57
SOLAR LIGHT 17
SOLENOIDS 117, 133
SOLID STATE RELAY 121
SOLID STATE RELAYS 120
SOLVENTS 12
SOURCES OF ERROR 43
SOUTH AMERICA 12
SOUTH SEAS TRADING
    COMPANY 22
SOUTHWEST AIR QUALITY
    DISTRICT 73
SPDT 193
SPECTRAL ANALYSES 16
SPECTRAL OUTPUT 180
SPECTRAL RESPONSE
    40, 161, 166
SPECTRAL SENSITIVITY
    147, 151
SPECTRAL SENSITIVITY CURVE
    106
SPECTRAL SIGNATURE 180
SPECTREX SHARPEYE 17
SPECTROPHOTOMETER 11
SPECTRUM 16, 168, 174, 175,
    176, 179, 180
SPEED 91, 92, 93, 97,
    104, 105, 110, 120, 140
SPOTLIGHT SENSOR 131
SQUARE WAVE 93, 185, 186,
    196
SQUAREWAVE OSCILLATOR 111
SQUEEGEE 14
STAMP COLLECTORS 9
STEFAN-BOLTZMANN'S
    EQUATION 32, 39
STN 76
STORAGE TIME 93
STRATOSPHERE 9, 10, 11, 12
SUBMODES 128
SUBNOTEBOOK COMPUTERS
    59
SUBSTITUTION ERROR RATE 69
SUBSTRATE 89
SUM FREQUENCY MIXING 10
SUMMING POINT 98
SUNSCREEN 22
SUNSHINE 15
SUNX 70
SUPPORT ELECTRONICS 40

SURFACE ACOUSTIC FILTER 57
SURFACE REFLECTIVITY 70, 71
SURFACE TEMPERATURE 36
SURFACE TEXTURE 36
SURFACES 44
SURGES 191
SWATCH 74
SWISS 75
SWITCHING SPEED 97, 98, 104,
    131
SYMBOL 69
SYMBOLOGY 69
SYMMETRY 106
SYNCHRONIZATION 122
SYNCHROTRONS 10

T

TANNING LOTION 22
TAUT BAND 188
TCP 80
TELEFUNKEN ICS 57
TEMPERATURE 18, 25, 26, 94,
    107, 118, 140, 179, 182
TEMPERATURE DIFFERENTIAL
    49
TEMPERATURE DRIFT 131
TEMPERATURE SCALES 33, 179
TEMPLE UNIVERSITY 18
TEST MODE 194
TEST PAGE 66
TEXAS INSTRUMENTS TSL230
    79
THE BEEP 74
THERMAL TEMPERATURE 174
THERMALERT 50
THERMO-DYNAMIC TRAP 49
THERMOMETER 25, 26, 32
THOMSON, WILLIAM (LORD
    KELVIN) 34, 179
THRESHOLD
    92, 108, 168, 169, 171
THRESHOLD ENERGY SIGNAL
    COMPARISON 16
THYRISTOR 109, 116, 122
TIMEX COMMUNICATION
    PROTOCOL 80
TIMEX DATALINK 76, 77
TIMING LOOP 165
TIRES 8
TITANIUM DIOXIDE 9

TMC3637 151
TONES 211
TRANSIENTS 119
TRANSISTOR 60, 89, 90, 92,
    93, 94, 97, 98, 104, 107,
    111, 122, 147, 148, 166
TRANSISTOR ARRAY PACK 214
TRANSITIONAL ELEMENTS 12
TRANSMITTER
    58, 146, 152, 166, 206, 212
TRANSPARENT OBJECT
    DETECTION 130
TRANSPARENT WINDOW 43
TRANSPONDER 57
TRANXIT SOFTWARE 62
TRAP OPERATION 47
TRAPS 48
TRIAC 109, 116, 117, 118, 119,
    120, 121, 123, 148,
    192, 200, 201
TRIAC DRIVER 97
TRIAC GATE 119, 200
TRIAC TEST 194
TRIATOMIC OXYGEN 11
TRIGGER POINT 97
TRIGGERED MODE 153
TRIGONOMETRIC REFLECTIVE
    PHOTOELECTRIC SENSOR
    131
TRIGONOMETRIC REFLECTIVE
    SENSOR 131, 140
TROPOSPHERE 10
TROUBLESHOOTING
    49, 186, 190
TSL230 150
TTL 200
TUNGSTEN FILAMENT 179
TUNGSTEN FILM 178
TUNGSTEN LAMP 179
TUNGSTEN LIGHT 178
TUTORIAL 174
TV 57
TV REMOTE CONTROLS 71, 83
TYPE A FILM 177

U

UART 81
ULTRASONIC WELDER 136
ULTRAVIOLET LIGHT 8, 179
UNIT SYMBOLS 20, 166
UNITY GAIN 104, 205

Index

UNIVERSAL ASYNCHRONOUS RECEIVER-TRANSMITTER 81
UNIVERSAL TIME 17
UPV 23
UV 8, 9, 10, 11, 12, 13, 14, 15, 16, 17, 18, 19, 20, 22, 139, 179, 190
UV BEAM ANALYSIS 25
UV CURING 14, 20
UV CURING SYSTEMS 18
UV DETECTORS 16, 17
UV EXPERIMENTAL INDEX 10
UV EXPOSING MECHANISM 9
UV FLAME DETECTORS 15, 16, 17
UV FLASHLAMP 21
UV GLOBAL NETWORK 18
UV INDEX 10
UV INTENSITY 18, 23, 25
UV LIGHT 8, 14, 18
UV METER 18
UV PHOTODIODE 22
UV RADIATION 11, 16, 18, 167, 190
UV RADIATION DETECTOR 174
UV SENSOR CARDS 25
UV SENSORS 15, 22, 23
UV SPECTRUM 16
UV-A 9
UV-B 9, 19, 190
UV-BIOMETER 17, 18
UV-C 9, 12, 22
UV-C R 21
UV-V 9, 19
UV/EB CURING 14
UVIMAP SYSTEM 18
UVTRON 22

V

VACUUM PHOTODIODE 19
VALANCE BAND 105
VALUE 176
VANTAGE NV 74
VCRS 57
VGA 76
VHF 75
VIEWING 43
VIRTUAL GROUND 98
VIRUSES 9
VISIBLE LIGHT 9, 22, 131, 166, 174, 213
VISIBLE LIGHT SPECTRUM 190
VISIBLE SPECTRUM 79, 180, 185, 190
VISUAL DATA 175
VOCS 13
VOICE MAIL 72
VOLATILE ORGANIC COMPOUNDS 11
VOLCANOES 12
VOLTAGE 88, 89, 92, 94, 97, 104, 109, 110, 116, 117, 118, 120, 121, 131, 134, 140, 147, 151, 166, 185, 186, 187, 190, 196, 205, 209
VOLTAGE GENERATOR 151
VOLTAGE REGULATOR 166, 184, 187, 188, 196, 206, 214
VOLTAGE SENSITIVITY 186
VOLTAGE SUPERVISORS 165
VOLTMETER 209
VON BEZOLD, WILHELM 176
VUV 9, 10

W

WATER VAPOR 17
WATT 20
WAVEFORM 98, 119, 164, 165, 168
WAVELENGTH 8, 9, 15, 17, 19, 26, 36, 37, 58, 70, 104, 105, 106, 110, 132, 175, 177
WESTON, EDWARD 188
WIEN'S DISPLACEMENT LAW 32, 39
WINDOWS 95 25
WINDOWS 3.1 25, 62
WINDOWS 95 62
WINDOWS FOR WORK GROUPS 62
WINDOWS NT, OS/2 WARP 62
WIRELESS CLIPBOARD 62
WORK 20
WRIST INSTRUMENTS 71, 72, 74, 75
WRIST REMOTE CONTROLLER 83
WRISTAPPS 78

X

X-RAYS 8

Z

ZERO CROSSING CURRENT 120
ZERO CROSSING DETECTOR 117, 122, 123, 192, 194
ZOOMER 56

The Howard W. Sams Troubleshooting & Repair Guide to TV
Howard W. Sams & Company

The Howard W. Sams Troubleshooting & Repair Guide to TV is the most complete and up-to-date television repair book available. Included in its more than 300 pages is complete repair information for all makes of TVs, timesaving features that even the pros don't know, comprehensive basic electronics information, and extensive coverage of common TV symptoms.

This repair guide is completely illustrated with useful photos, schematics, graphs, and flowcharts. It covers audio, video, technician safety, test equipment, power supplies, picture-in-picture, and much more. *The Howard W. Sams Troubleshooting & Repair Guide to TV* was written, illustrated, and assembled by the engineers and technicians of Howard W. Sams & Company. This book is the first truly comprehensive television repair guide published in the 90s, and it contains vast amounts of information never printed in book form before.

Video Technology
384 pages ◆ Paperback ◆ 8-1/2 x 11"
ISBN: 0-7906-1077-9 ◆ Sams: 61077
$29.95 ($39.95 Canada) ◆ June 1996

The In-Home VCR Mechanical Repair & Cleaning Guide
Curt Reeder

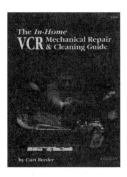

Like any machine that is used in the home or office, a VCR requires minimal service to keep it functioning well and for a long time. However, a technical or electrical engineering degree is not required to begin regular maintenance on a VCR. *The In-Home VCR Mechanical Repair & Cleaning Guide* shows readers the tricks and secrets of VCR maintenance using just a few small hand tools, such as tweezers and a power screwdriver.

This book is also geared toward entrepreneurs who may consider starting a new VCR service business of their own. The vast information contained in this guide gives a firm foundation on which to create a personal niche in this unique service business. This book is compiled from the most frequent VCR malfunctions Curt Reeder has encountered in the six years he has operated his in-home VCR repair and cleaning service.

Video Technology
222 pages ◆ Paperback ◆ 8-3/8 x 10-7/8"
ISBN: 0-7906-1076-0 ◆ Sams: 61076
$19.95 ($26.99 Canada) ◆ April 1996

ES&T Presents TV Troubleshooting & Repair
Electronic Servicing & Technology Magazine

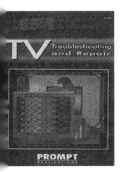

TV set servicing has never been easy. The service manager, service technician, and electronics hobbyist need timely, insightful information in order to locate the correct service literature, make a quick diagnosis, obtain the correct replacement components, complete the repair, and get the TV back to the owner.

ES&T Presents TV Troubleshooting & Repair presents information that will make it possible for technicians and electronics hobbyists to service TVs faster, more efficiently, and more economically, thus making it more likely that customers will choose not to discard their faulty products, but to have them restored to service by a trained, competent professional.

Originally published in *Electronic Servicing & Technology*, the chapters in this book are articles written by professional technicians, most of whom service TV sets every day. These chapters provide general descriptions of television circuit operation, detailed service procedures, and diagnostic hints.

Video Technology
226 pages ◆ Paperback ◆ 6 x 9"
ISBN: 0-7906-1086-8 ◆ Sams: 61086
$18.95 ($25.95 Canada) ◆ August 1996

CALL 1-800-428-7267 TODAY FOR THE NAME OF YOUR NEAREST PROMPT PUBLICATIONS DISTRIBUTOR

Theory & Design of Loudspeaker Enclosures
Dr. J. Ernest Benson

The design of loudspeaker enclosures, particularly vented enclosures, has been a subject of continuing interest since 1930. Since that time, a wide range of interests surrounding loudspeaker enclosures have sprung up that grapple with the various aspects of the subject, especially design. *Theory & Design of Loudspeaker Enclosures* lays the groundwork for readers who want to understand the general functions of loudspeaker enclosure systems and eventually experiment with their own design.

Written for design engineers and technicians, students and intermediate-to-advanced level acoustics enthusiasts, this book presents a general theory of loudspeaker enclosure systems. Full of illustrated and numerical examples, this book examines diverse developments in enclosure design, and studies the various types of enclosures as well as varying parameter values and performance optimization.

Audio Technology
244 pages ◆ Paperback ◆ 6 x 9"
ISBN: 0-7906-1093-0 ◆ Sams: 61093

TV Video Systems
L.W. Pena & Brent A. Pena

Knowing which video programming source to choose, and knowing what to do with it once you have it, can seem overwhelming. Covering standard hard-wire cable, large-dish satellite systems, and DSS, *TV Video Systems* explains the different systems, how they are installed, their advantages and disadvantages, and how to troubleshoot problems. This book presents easy-to-understand information and illustrations covering installation instructions, home options, apartment options, detecting and repairing problems, and more. The in-depth chapters guide you through your TV video project to a successful conclusion.

L.W. Pena is an independent certified cable TV technician with 14 years of experience who has installed thousands of TV video systems in homes and businesses. Brent Pena has eight years of experience in computer science and telecommunications, with additional experience as a cable installer.

Video Technology
124 pages ◆ Paperback ◆ 6 x 9"
ISBN: 0-7906-1082-5 ◆ Sams: 61082

The Video Book
Gordon McComb

Televisions and video cassette recorders have become part of everyday life, but few people know how to get the most out of these home entertainment devices. *The Video Book* offers easy-to-read text and clearly illustrated examples to guide readers through the use, installation, connection, and care of video system components. Simple enough for the new buyer, yet detailed enough to assure proper connection of the units after purchase, this book is a necessary addition to the library of every modern video consumer. Topics included in the coverage are the operating basics of TVs, VCRs, satellite systems, and video cameras; maintenance and troubleshooting; and connectors, cables, and system interconnections.

Gordon McComb has written over 35 books and 1,000 magazine articles, which have appeared in such publications as *Popular Science*, *Video*, *PC World*, and *Omni*, as well as many other top consumer and trade publications. His writing has spanned a wide range of subjects, from computers to video to robots.

Video Technology
192 pages ◆ Paperback ◆ 6 x 9"
ISBN: 0-7906-1030-2 ◆ Sams: 61030
$16.95 ($22.99 Canada) ◆ October 1992

CALL 1-800-428-7267 TODAY FOR THE NAME OF YOUR NEAREST PROMPT PUBLICATIONS DISTRIBUTOR

Is This Thing On?
Gordon McComb

Is This Thing On? takes readers through each step of selecting components, installing, adjusting, and maintaining a sound system for small meeting rooms, churches, lecture halls, public-address systems for schools or offices, or any other large room.

In easy-to-understand terms, drawings and illustrations, *Is This Thing On?* explains the exact procedures behind connections and troubleshooting diagnostics. With the help of this book, hobbyists and technicians can avoid problems that often occur while setting up sound systems for events and lectures.

Is This Thing On? covers basic components of sound systems, the science of acoustics, enclosed room, sound system specifications, wiring sound systems, and how to install wireless microphones, CD players, portable public-address systems, and more.

Audio Technology
136 pages ♦ Paperback ♦ 6 x 9"
ISBN: 0-7906-1081-7 ♦ Sams: 61081
$14.95 ($20.95 Canada) ♦ April 1996

Advanced Speaker Designs
Ray Alden

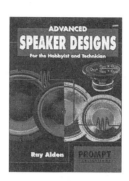

Advanced Speaker Designs shows the hobbyist and the experienced technician how to create high-quality speaker systems for the home, office, or auditorium. Every part of the system is covered in detail, from the driver and crossover network to the enclosure itself. Readers can build speaker systems from the parts lists and instructions provided, or they can actually learn to calculate design parameters, system responses, and component values with scientific calculators or PC software.

This book includes construction plans for seven complete systems, easy-to-understand instructions and illustrations, and chapters on sealed and vented enclosures. There is also emphasis placed on enhanced bass response, computer-aided speaker design, and driver parameters. *Advanced Speaker Designs* is a companion book to *Speakers for Your Home and Automobile*, also available from Prompt® Publications.

Audio Technology
136 pages ♦ Paperback ♦ 6 x 9"
ISBN: 0-7906-1070-1 ♦ Sams: 61070
$16.95 ($22.99 Canada) ♦ July 1995

Making Sense of Sound
Alvis J. Evans

This book deals with the subject of sound — how it is detected and processed using electronics in equipment that spans the full spectrum of consumer electronics. It concentrates on explaining basic concepts and fundamentals to provide easy-to-understand information, yet it contains enough detail to be of high interest to the serious practitioner. Discussion begins with how sound propagates and common sound characteristics, before moving on to the more advanced concepts of amplification and distortion. *Making Sense of Sound* was designed to cover a broad scope, yet in enough detail to be a useful reference for readers at every level.

Alvis Evans is the author of many books on the subject of electricity and electronics for beginning hobbyists and advanced technicians. He teaches seminars and workshops worldwide to members of the trade, as well as being an Associate Professor of Electronics at Tarrant County Junior College.

Audio Technology
112 pages ♦ Paperback ♦ 6 x 9"
ISBN: 0-7906-1026-4 ♦ Sams: 61026
$10.95 ($14.95 Canada) ♦ November 1992

CALL 1-800-428-7267 TODAY FOR THE NAME OF YOUR NEAREST PROMPT PUBLICATIONS DISTRIBUTOR

Semiconductor Cross Reference Book
Fourth Edition
Howard W. Sams & Company

This newly revised and updated reference book is the most comprehensive guide to replacement data available for engineers, technicians, and those who work with semiconductors. With more than 490,000 part numbers, type numbers, and other identifying numbers listed, technicians will have no problem locating the replacement or substitution information needed. There is not another book on the market that can rival the breadth and reliability of information available in the fourth edition of the *Semiconductor Cross Reference Book*.

Professional Reference
688 pages ◆ Paperback ◆ 8-1/2 x 11"
ISBN: 0-7906-1080-9 ◆ Sams: 61080
$24.95 ($33.95 Canada) ◆ August 1996

The Component Identifier and Source Book
Victor Meeldijk

Because interface designs are often reverse engineered using component data or block diagrams that list only part number technicians are often forced to search for replacement parts armed only with manufacturer logos and part numbers.

This source book was written to assist technicians and system designers in identifying components from prefixes and logos, as well as find sources for various types of microcircuits and other components. There is not another book on the market that lists as many manufacturers of such diverse electronic components.

Professional Reference
384 pages ◆ Paperback ◆ 8-1/2 x 11"
ISBN: 0-7906-1088-4 ◆ Sams: 61088
$24.95 ($33.95 Canada) ◆ November 1996

IC Cross Reference Book
Second Edition
Howard W. Sams & Company

The engineering staff of Howard W. Sams & Company assembled the *IC Cross Reference Book* to help readers find replacements or substitutions for more than 35,000 ICs and modules. It is an easy-to-use cross reference guide and includes part numbers for the United States, Europe, and the Far East. This reference book was compiled from manufacturers' data and from the analysis of consumer electronics devices for PHOTOFACT® service data, which has been relied upon since 1946 by service technicians worldwide.

Professional Reference
192 pages ◆ Paperback ◆ 8-1/2 x 11"
ISBN: 0-7906-1096-5 ◆ Sams: 61096
$19.95 ($26.99 Canada) ◆ November 1996

Tube Substitution Handbook
William Smith & Barry Buchanan

The most accurate, up-to-date guide available, the *Tube Substitution Handbook* is useful to antique radio buffs, old car enthusiasts, and collectors of vintage ham radio equipment. In addition, marine operators, microwave repair technicians, and TV and radio technicians will find the *Handbook* to be an invaluable reference tool.

The *Tube Substitution Handbook* divided into three sections, each preceded by specific instructions. These sections are vacuum tubes, picture tubes, and tube basing diagrams.

Professional Reference
149 pages ◆ Paperback ◆ 6 x 9"
ISBN: 0-7906-1036-1 ◆ Sams: 61036
$16.95 ($22.99 Canada) ◆ March 1995

CALL 1-800-428-7267 TODAY FOR THE NAME OF YOUR NEAREST PROMPT PUBLICATIONS DISTRIBUTOR

Alternative Energy
Mark E. Hazen

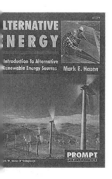

This book is designed to introduce readers to the many different forms of energy mankind has learned to put to use. Generally, energy sources are harnessed for the purpose of producing electricity. This process relies on transducers to transform energy from one form into another. *Alternative Energy* will not only address transducers and the five most common sources of energy that can be converted to electricity, it will also explore solar energy, the harnessing of the wind for energy, geothermal energy, and nuclear energy.

This book is designed to be an introduction to energy and alternate sources of electricity. Each of the nine chapters are followed by questions to test comprehension, making it ideal for students and teachers alike. In addition, listings of World Wide Web sites are included so that readers can learn more about alternative energy and the organizations devoted to it.

Professional Reference
320 pages ♦ Paperback ♦ 7-3/8 x 9-1/4"
ISBN: 0-7906-1079-5 ♦ Sams: 61079
$18.95 ($25.95 Canada) ♦ October 1996

The Complete RF Technician's Handbook
Cotter W. Sayre

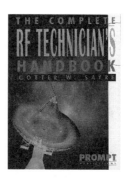

The *Complete RF Technician's Handbook* will furnish the working technician or student with a solid grounding in the latest methods and circuits employed in today's RF communications gear. It will also give readers the ability to test and troubleshoot transmitters, transceivers, and receivers with absolute confidence. Some of the topics covered include reactance, phase angle, logarithms, diodes, passive filters, amplifiers, and distortion. Various multiplexing methods and data, satellite, spread spectrum, cellular, and microwave communication technologies are discussed.

Cotter W. Sayre is an electronics design engineer with Goldstar Development, Inc., in Lake Elsinore, California. He is a graduate of Los Angeles Pierce College and is certified by the National Association of Radio and Telecommunications Engineers, as well as the International Society of Electronics Technicians.

Professional Reference
281 pages ♦ Paperback ♦ 8-1/2 x 11"
ISBN: 0-7906-1085-X ♦ Sams: 61085
$24.95 ($33.95 Canada) ♦ July 1996

Surface-Mount Technology for PC Boards
James K. Hollomon, Jr.

The race to adopt surface-mount technology, or SMT as it is known, has been described as the latest revolution in electronics. This book is intended for the working engineer or manager, the student or the interested layman, who would like to learn to deal effectively with the many trade-offs required to produce high manufacturing yields, low test costs, and manufacturable designs using SMT. The valuable information presented in *Surface-Mount Technology for PC Boards* includes the benefits and limitations of SMT, SMT and FPT components, manufacturing methods, reliability and quality assurance, and practical applications.

James K. Hollomon, Jr. is the founder and president of AMTI, an R&D and prototyping service concentrating on miniaturization and low-noise, high-speed applications. He has nearly 20 years experience in engineering, marketing, and managing firms dealing with leadless components. His previous appointments include national president of the Surface-Mount Technology Association.

Professional Reference
510 pages ♦ Paperback ♦ 7 x 10"
ISBN: 0-7906-1060-4 ♦ Sams: 61060
$26.95 ($36.95 Canada) ♦ July 1995

CALL 1-800-428-7267 TODAY FOR THE NAME OF YOUR NEAREST PROMPT PUBLICATIONS DISTRIBUTOR

Internet Guide to the Electronics Industry
John Adams

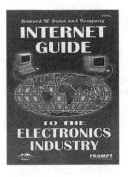

Although the Internet pervades our lives, it would not have been possible without the growth of electronics. It is very fitting then that technical subjects, data sheets, parts houses, and of course manufacturers, are developing new and innovative ways to ride along the Information Superhighway. Whether it's programs that calculate Ohm's Law or a schematic of a satellite system, electronics hobbyists and technicians can find a wealth of knowledge and information on the Internet.

In fact, soon electronics hobbyists and professionals will be able to access on-line catalogs from manufacturers and distributors all over the world, and then order parts, schematics, and other merchandise without leaving home. The *Internet Guide to the Electronics Industry* serves mainly as a directory to the resources available to electronics professionals and hobbyists.

Internet
192 pages ♦ Paperback ♦ 5-1/2 x 8-1/2"
ISBN: 0-7906-1092-2 ♦ Sams: 61092
$16.95 ($22.99 Canada) ♦ December 1996

Real-World Interfacing with Your PC
James "J.J." Barbarello

As the computer becomes increasingly prevalent in society, its functions and applications continue to expand. Modern software allows users to do everything from balance a checkbook to create a family tree. Interfacing, however, is truly the wave of the future for those who want to use their computer for things other than manipulating text, data, and graphics.

Real-World Interfacing With Your PC provides all the information necessary to use a PC's parallel port as a gateway to electronic interfacing. In addition to hardware fundamentals, this book provides a basic understanding of how to write software to control hardware.

While the book is geared toward electronics hobbyists, it includes a chapter on project design and construction techniques, a checklist for easy reference, and a recommended inventory of starter electronic parts to which readers at every level can relate.

Computer Technology
119 pages ♦ Paperback ♦ 7-3/8 x 9-1/4"
ISBN: 0-7906-1078-7 ♦ Sams: 61078
$16.95 ($22.99 Canada) ♦ March 1996

ES&T Presents Computer Troubleshooting & Repair
Electronic Servicing & Technology

ES&T is the nation's most popular magazine for professionals who service consumer electronics equipment. PROMPT Publications, a rising star in the technical publishing business, is combining its publishing expertise with the experience and knowledge of *ES&T's* best writers to produce a new line of troubleshooting and repair books for the electronic market. Compiled from articles and prefaced by the editor in chief, Nils Conrad Persson, these books provide valuable hands-on information for anyone interested in electronics and product repair.

Computer Troubleshooting & Repair is the second book in the series and features information on repairing Macintosh computers, a CD-ROM primer, and a color monitor. Also included are hard drive troubleshooting and repair tips, computer diagnostic software, networking basics, preventative maintenance for computers, upgrading, and much more.

Computer Technology
288 pages ♦ Paperback ♦ 6 x 9"
ISBN: 0-7906-1087-6 ♦ Sams: 61087
$18.95 ($26.50 Canada) ♦ February 1997

**CALL 1-800-428-7267 TODAY FOR THE NAME OF
YOUR NEAREST PROMPT PUBLICATIONS DISTRIBUTOR**

The Phone Book
Gerald Luecke & James Allen

This book is an installation guide for telephones and telephone accessories. It was written to make it easier for the inexperienced person to install telephones, whether existing ones are being replaced or moved or new ones added, without the hassle and expense of contracting a serviceman. *The Phone Book* begins by explaining the telephone system and its operation, before moving onto clear step-by-step instructions for replacing and adding telephones. With this book, a minimum of tools available around the house, and readily available parts, readers will be able to handle any telephone installation in the home, apartment, or small business.

Gerald Luecke has written articles on integrated circuits and digital technology for numerous trade and professional organizations. James Allen is the President, CEO, and a director of Master Publishing.

Communication
176 pages ♦ Paperback ♦ 7-3/8 x 9-1/4"
ISBN: 0-7906-1028-0 ♦ Sams: 61028
$16.95 ($22.99 Canada) ♦ October 1992

Digital Electronics
Stephen Kamichik

Although the field of digital electronics emerged years ago, there has never been a definitive guide to its theories, principles, and practices — until now. *Digital Electronics* is written as a textbook for a first course in digital electronics, but its applications are varied.

Useful as a guide for independent study, the book also serves as a review for practicing technicians and engineers. And because *Digital Electronics* does not assume prior knowledge of the field, the hobbyist can gain insight about digital electronics.

Some of the topics covered include analog circuits, logic gates, flip-flops, and counters. In addition, a problem set appears at the end of each chapter to test the reader's understanding and comprehension of the materials presented. Detailed instructions are provided so that the readers can build the circuits described in this book to verify their operation.

Electronic Theory
150 pages ♦ Paperback ♦ 7-3/8 x 9-1/4"
ISBN: 0-7906-1075-2 ♦ Sams: 61075
$16.95 ($22.99 Canada) ♦ February 1996

The Right Antenna
Alvis J. Evans

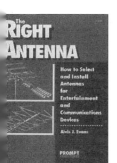

The Right Antenna is intended to provide easy-to-understand information on a wide variety of antennas. It begins by explaining how antennas work and then isolates antennas for TV and FM. A separate chapter is devoted to satellite TV antennas, noise and interference, and antennas used by hams for antenna band operation. The basic concepts of cellular telephone system operation are explained and the most popular antennas are discussed. After studying this book, the reader will be able to select an antenna, place it correctly, and install it properly to obtain maximum performance whether in a strong signal area or in a fringe area.

Alvis Evans is the author of many books on the subject of electricity and electronics for beginning hobbyists and advanced technicians. He teaches seminars and workshops worldwide to members of the trade, as well as being an Associate Professor of Electronics at Tarrant County Junior College.

Communication
112 pages ♦ Paperback ♦ 6 x 9"
ISBN: 0-7906-1022-1 ♦ Sams: 61022
$10.95 ($14.95 Canada) ♦ November 1992

CALL 1-800-428-7267 TODAY FOR THE NAME OF YOUR NEAREST PROMPT PUBLICATIONS DISTRIBUTOR

Semiconductor Essentials
Stephen Kamichik

Readers will gain hands-on knowledge of semiconductor diodes and transistors with help from the information in this book. *Semiconductor Essentials* is a first course in electronics at the technical and engineering levels. Each chapter is a lesson in electronics, with problems included to test understanding of the material presented. This generously illustrated manual is a useful instructional tool for the student and hobbyist, as well as a practical review for professional technicians and engineers. The comprehensive coverage includes semiconductor chemistry, rectifier diodes, zener diodes, transistor biasing, and more.

Author Stephen Kamichik is an electronics consultant who has developed dozens of electronic products and received patents in both the U.S. and Canada. He holds degrees in electrical engineering, and was employed by SPAR, where he worked on the initial prototyping of the Canadarm.

Electronic Theory
112 pages ♦ Paperback ♦ 6 x 9"
ISBN: 0-7906-1071-X ♦ Sams: 61071
$16.95 ($22.99 Canada) ♦ September 1995

Introduction to Microprocessor Theory & Operation
J.A. Sam Wilson & Joseph Risse

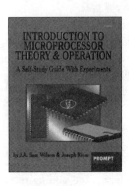

This book takes readers into the heart of computerized equipment and reveal how microprocessors work. By covering digital circuits in addition to microprocessors and providing self-tests and experiments, *Introduction to Microprocessor Theory & Operation* makes it easy to learn microprocessor systems. The text is full illustrated with circuits, specifications, an pinouts to guide beginners through the ins and-outs of microprocessors, as well a provide experienced technicians with a valuable reference and refresher tool.

J.A. Sam Wilson has written numerous books covering all aspects of the electronics field, and has served as the Director of Technical Publications for NESDA. Joseph Risse develops courses an laboratory experiments in self-study and industrial electronics for International Correspondence Schools and other independent study schools.

Electronic Theory
211 pages ♦ Paperback ♦ 6 x 9"
ISBN: 0-7906-1064-7 ♦ Sams: 61064
$16.95 ($22.99 Canada) ♦ February 1995

Basic Principles of Semiconductors
Irving M. Gottlieb

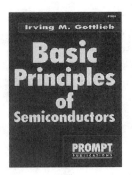

Despite their ever-growing prominence in the electronics industry, semiconductors are still plagued by a stigma which defines them merely as poor conductors. This narrow-sighted view fails to take into account the fact that semiconducto are truly unique alloys whose conductivity is enhanced tenfold by the addition of even the smallest amount of ligh voltage, heat, or certain other substances. *Basic Principles of Semiconductors* explores the world of semiconductor beginning with an introduction to atomic physics before moving onto the structure, theory, applications, and future these still-evolving alloys. Such a theme makes this book useful to a wide spectrum of practitioners, from the hobbyi and student, right up to the technician and the professional electrician. Irving M. Gottlieb is the author of over ten bool in the electrical and electronics fields. *Basic Principles of Semiconductors* is his latest offering, however *Test Procedur for Basic Electronics* is also available from PROMPT® Publications.

Electronic Theory
158 pages ♦ Paperback ♦ 6 x 9"
ISBN: 0-7906-1066-3 ♦ Sams: 61066
$14.95 ($20.95 Canada) ♦ April 1995

CALL 1-800-428-7267 TODAY FOR THE NAME OF YOUR NEAREST PROMPT PUBLICATIONS DISTRIBUTOR

Managing the Computer Power Environment
Mark Waller

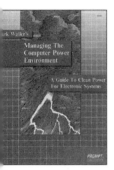

Clean power is what every computer system needs to operate without error. But electricity's voyage from utility company to home or office introduces noise, surges, static, and a host of gremlins that can seriously affect computer performance and data security. Written for data processing specialists, field engineers, technicians, and computer network professionals, *Managing the Computer Power Environment* provides the background in electrical technology that will help readers understand and control the quality of the power that drives their computer system. Covering utility power, grounding, power distribution units, and backup power systems and conditioners, Mark Waller prepares the reader to manage the demons of electrical destruction through ensuring clean power for your electronic system. Mark Waller is president of the Waller Group, Inc., a company specializing in solving electrical power and grounding problems.

Electrical Technology
174 pages ◆ Paperback ◆ 7-3/8 x 9-1/4"
ISBN: 0-7906-1020-5 ◆ Sams: 61020
$19.95 ($26.99 Canada) ◆ April 1992

Surges, Sags and Spikes
Mark Waller

Surges, sags, spikes, brownouts, blackouts, lightning and other damaging electrical power disturbances can render a personal computer system and its data useless in a few milliseconds — unless you're prepared. Mark Waller's *Surges, Sags and Spikes* is written for all personal computer users concerned with protecting their computer systems against a hostile electrical environment. In easy-to-understand, non-technical language, the author takes a comprehensive look at approaches to solving computer power problems. Helpful diagrams and photographs are included to document computer power needs and solutions.

Mark Waller is an award-winning author whose numerous articles have appeared in such magazines as *Byte*, *Datamation*, and *Network World*. He is the author of another book dealing with power entitled *Managing the Computer Power Environment*, also available from PROMPT® Publications.

Electrical Technology
220 pages ◆ Paperback ◆ 7-3/8 x 9-1/4"
ISBN: 0-7906-1019-1 ◆ Sams: 61019
$19.95 ($26.99 Canada) ◆ April 1992

Harmonics
Mark Waller

Harmonics is the essential guide to understanding all of the issues and areas of concern surrounding harmonics and the recognized methods for dealing with them. Covering nonlinear loads, multiple PCs, K-factor transformers, and more, Mark Waller prepares the reader to manage problems often encountered in electrical distribution systems that can be solved easily through an understanding of harmonics, current, and voltage. This book is a useful tool for system and building engineers, electricians, maintenance personnel, and all others concerned about protecting and maintaining the quality of electrical power systems.

Mark Waller is president of the Waller Group, Inc., and specializes in harmonic analysis and in solving electrical power and grounding problems for facilities. He has been actively involved in the field of electrical power quality for many years. Waller has a broad background in all aspects of power quality, power protection, and system integrity.

Electrical Technology
132 pages ◆ Paperback ◆ 7-3/8 x 9-1/4"
ISBN: 0-7906-1048-5 ◆ Sams: 61048
$24.95 ($33.95 Canada) ◆ May 1994

CALL 1-800-428-7267 TODAY FOR THE NAME OF YOUR NEAREST PROMPT PUBLICATIONS DISTRIBUTOR

COMING IN 1997 —
Optoelectronics
Volumes 1 and 3 !

COMING Winter 1997!
Optoelectronics
Volume 1
Vaughn D. Martin

Optoelectronics, Volume 1 is an introductory self-teaching text that contains:

- Basic Optoelectronic Concepts!
- Photometrics!
- Optics!
- AND MUCH MORE!

Optoelectronics, Volume 1
286 pages ♦ Paperback ♦ 8-1/2 x 11"
ISBN: 0-7906-1091-4 ♦ Sams: 61091
$29.95 ♦ January 1997

CALL 1-800-428-7267 TODAY FOR THE NAME OF YOUR NEAREST PROMPT PUBLICATIONS DISTRIBUTOR

About The Author

Vaughn D. Martin is a senior electrical engineer with the Department of the Air Force. Previously he worked at Magnavox and ITT Aerospace/Optics, where he acquired his fascination with optoelectronics. He has published numerous articles in trade, amateur radio, electronic hobbyist, troubleshooting and repair, and optoelectronics magazines. He has also written several books covering a wide range of topics in the field of electronics.

COMING SUMMER 1997!
Optoelectronics
Volume 3
Vaughn D. Martin

Optoelectronics, Volume 3 is an Advanced self-teaching text that contains:

- Information on Fiber Optics!

- Lab experiments for your own home lab, professional lab, or classroom!

- AND MUCH MORE!

Optoelectronics, Volume 3
400 pages ◆ Paperback ◆ 8-1/2 x 11"
ISBN: PENDING ◆ Sams: PENDING
$29.95 ◆ August 1997

**CALL 1-800-428-7267 TODAY FOR THE NAME OF
YOUR NEAREST PROMPT PUBLICATIONS DISTRIBUTOR**

Howard W. Sams
A Bell Atlantic Company

Your Technology Connection to the Future!

Now You Can Visit

Howard W. Sams & Company <u>On-Line</u>:
http://www.hwsams.com

Gain Easy Access to:

- **The PROMPT Publications catalog, for information on our Latest Book Releases.**
- **The PHOTOFACT Annual Index.**
- **Information on Howard W. Sams' Latest Products.**
- **AND MORE!**

CALL 1-800-428-7267 TODAY FOR THE NAME OF YOUR NEAREST PROMPT PUBLICATIONS DISTRIBUTOR